APE MAN

APE MAN

THE STORY OF HUMAN EVOLUTION

ROD CAIRD

Consultant Scientific Editor
DR ROBERT FOLEY

Consultant For Illustrations: Dr Stephen Gooder

B⬤XTREE

For
Charlie, Jack and Luke –
a new generation

First published in Great Britain in 1994 by Boxtree Limited
Introduction © Dr Robert Foley 1994
Main text © Rod Caird 1994
Illustrations pages 26, 30, 34, 45 above and below, 49, 56, 60, 77, 80, 86, 93, 101, 105, 114, 119,
120, 121, 129, 136 © Stephen Gooder 1994
Illustrations pages 16/17, 24/25, 46/47, 62/63, 74/75, 82/83, 102/103, 138/139, 142/143,
154/155 © Christa Hook 1994
The right of Rod Caird to be identified as Author of this Work has been asserted by him in
accordance with the Copyright, Designs and Patents Act 1988.

Designed by Robert Updegraff
Printed and bound in the UK by Bath Press Colour Books, Glasgow for
Boxtree Limited
Broadwall House
21 Broadwall
London SE1 9PL

A CIP catalogue entry for this book is available from the British Library.
ISBN 1 85283 424 2

Based on the television series "Ape Man" produced by Granada Television

 GRANADA TELEVISION

ARTS & ENTERTAINMENT NETWORK

Contents

PREFACE

Some years ago I was responsible for a series of television programs called *Dinosaur!* in which Walter Cronkite told the story of the most successful creatures which have ever walked the earth. In four hours, we covered about 150 million years of history, and when the Arts & Entertainment channel in the US and ITV in the UK broadcast the programs, they seemed to touch a nerve.

It would be good, if optimistic, to think that *Jurassic Park* followed our lead. In fact it was nothing but coincidence and luck that our series was seen at a time when the whole world was caught up with the fascination of the dinosaurs.

The distant past is a wonderful subject. People seem to have a limitless appetite for learning about it, possibly because of the insecurities of the present. It may be that one of the magnetic qualities of the story of the dinosaurs, apart from the fact that the creatures themselves are so extraordinary, is that it puts our own existence as humans into clear perspective. By any measure, there have been humans on this planet for only a maximum of seven million years – and this in the context of a history of life of all kinds which stretches back more than three billion years.

After the dinosaurs, what next? When they became extinct, somewhere around sixty million years ago, they left behind their successor class of creatures, the mammals. And from among the mammals emerged, by some remarkable process of change, the first upright-walking apes, the distant ancestors of today's humans.

At its simplest, this is the story of who we are. Why do humans exist? Why are we the way we are? What was life like for our earliest ancestors? What may the future hold for us?

Life for humans, as we approach the end of the twentieth century, is not easy. The world is a troubled place. What messages can the past offer us about the present?

Our new television series, again with Walter Cronkite as guide and story-teller, tries to answer some of these questions, and this book offers an opportunity to be more expansive in exploring them.

It is a book about ideas. In the past twenty-five years, a good deal of the factual framework of human evolution has been established with a greater degree of certainty than ever before, and today's evolutionary scientists are increasingly preoccupied with seeking explanations as well as with adding to the known facts of our history.

There have been some great scientific advances in recent times. Gone are the days when whole theories about the past were constructed on the basis of a single fossil fragment. Now the continued search for actual human remains, and for the rubbish early humans left behind, takes its place alongside the complementary study of genetics, the climate, the environment and other animals in building up a picture of the past.

Gone are the days, too, when the history of humanity was seen as an upward march from the primitive to the civilized, when the past was peopled with creatures whose only virtue was that one day they would learn to be proper humans.

Ape Man

Walter Cronkite on location in Kenya for the filming of Ape Man *in the summer of 1993.*

Modern science explores why and when humans and our ancestors acquired the characteristics which we like to think make us so special – upright walking, the use of tools and technology, language, imagination and creativity – and sets them in the context of time. Without giving away too much of the underlying story at this early stage, it is worth just pausing to think over the idea that the time we are living in now is very probably the first period during which there has been only one species of humans alive. And the idea that extinction is an iron rule of biology.

This book follows a thematic approach. It tries to pick apart the species of *Homo sapiens* and to explain why the search for the origins of humanity will never come up with a single answer. As one scientist says: "It all depends what you mean by human." It tries to disentangle the fact that we think of ourselves as unique from the fact that all species of life are unique; what is it about us that makes us think we are even more special?

This is a layman's book. I am not a scientist, and I have therefore relied on the unstinting help given while we were making the programs by academics all over the world. Without their generosity with their time and their ideas, the programs and the book would have been impossible. Many of them are directly quoted in the book, from interviews they gave members of the production team when we were filming. I would like to take this opportunity to thank each of them, most sincerely, for their willingness and patience.

In particular I would like to thank Robert Foley of Cambridge University, who acted as scientific consultant on both the television programs and this book. He has been endlessly calm and tolerant in explaining facts and ideas which to him are utterly elementary, and in answering questions which even I can see, after a year's exposure to the subject, have occasionally been hopelessly naive. His contribution, on and off the screen, has been indispensable.

I also want to thank Stephen Gooder, who made sure that all Christa Hook's fine illustrations in the program and the book were scientifically sound, and who has himself designed many of the illustrations and graphics in the book. Susanna Wadeson at Boxtree Ltd has been an extremely patient editor and an invaluable critic and supporter.

Television programs are a collaborative effort, and the book would not have happened without the huge contribution of research and ideas from the entire production team on the series. My special thanks are therefore due to Jim Black and Peter Nicholson (directors), Jemma Jupp (associate producer), Becky Jones (researcher), Andrea Sprunt (line producer) and Samantha Stuart (production secretary), without any one of whom none of this would have been possible. The programs were also very fortunate to benefit from the creative skills of Peter Terry and Matt Howarth (graphics), Roger Bolton (music) and Kim Horton and Tony Cranstoun (editors). Michael Katz at the Arts & Entertainment Network had what I hope will turn out to be the foresight to commission a series of programs on a subject which by definition is not there to be filmed, and his nerve has never failed.

Responsibility for what follows, however, is mine.

ROD CAIRD
LONDON
FEBRUARY 1994

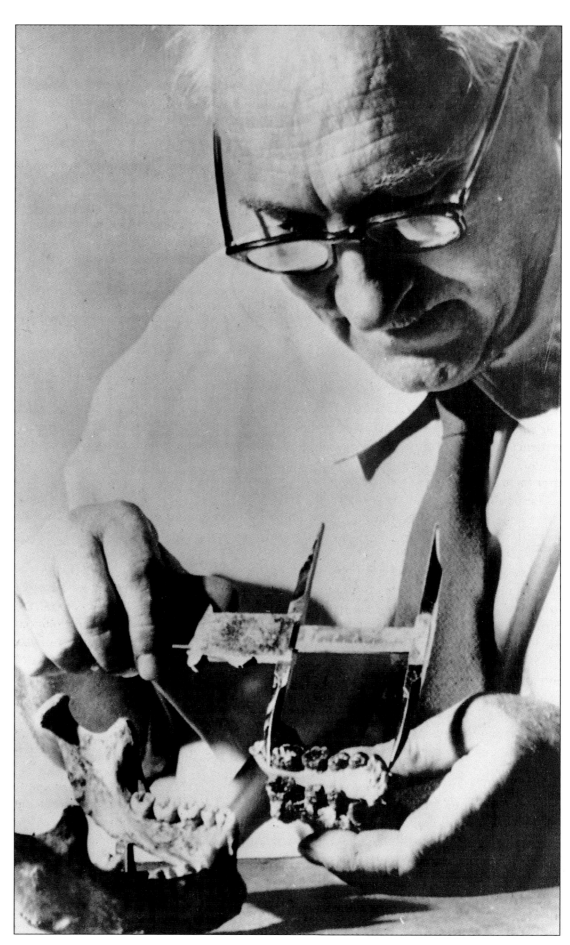

Louis Leakey (1903-72), one of the great individualists of fossil hunting, can take credit together with his wife Mary and their son Richard for establishing the importance of East Africa in the evolutionary story.

Introduction

Human evolution is all things to all people. To some it is a creed fervently to be believed, the victory of twentieth-century rationalism over primitive superstition. To others it is an exciting and romantic adventure, the search for fossils that trace our ancestry, and the hope that somewhere out there is the Holy Grail of the missing link between humans and apes.

To others still it is a lesson in human frailty and subjectivity, as the story of our search for our ancestors is replete with occasions where belief and prejudice have triumphed over hard evidence. Indeed to some it may appear as a story of frauds and forgers, and disputes of almost epic scale. To some, too, it is the ultimate evil, the heretical ascendancy of science over the truth of the Bible.

Certainly the story is all these things, much more, and in some ways much less. After all, at heart it is really just a dull science made up of a few dirty and broken fossils that, ironically, just might provide the answer to life, the universe and everything.

Evolution is also, of course, a highly specialized and technical subject. There are the palaeontologists who seek out the fossils, the geologists without whom no one would know where to look, the archaeologists who can put the flesh of behavior on to the bare bones of long-gone peoples, the chemists and physicists who can date the fossils and work out the environments of the past; and, increasingly, the geneticists who can show how our evolutionary history is found in each of the cells of our bodies as much as in the ground. The fact that what we know about human evolution is dependent upon the haphazard and rare process of fossilization – the preservation of bones and teeth after death – has led to the not unreasonable conclusion that palaeoanthropology is a fragile science. Each new fossil can totally rewrite the story of human evolution. Our species can go back a few hundred thousand or even million years overnight. One day we have our origins in Europe, the next in Africa. This is not the certainty that we generally expect from science. The newspaper headlines often gloat about the overthrow of yet another theory of human evolution.

While it will always be true that the fossil record of our history is partial and incomplete, nonetheless the evidence is growing all the time. When Raymond Dart found the Taung baby, the first australopithecine, it was the first fossil from Africa. When Mary Leakey discovered the fossil remains of "Nutcracker Man" in Olduvai Gorge it was the first known early hominid from East Africa, and one of only a handful of specimens from the continent as a whole. Now if a new fossil turns up it is just one among thousands that are known. This means that the significance of each new fossil diminishes, and its potential for overthrowing all the other evidence is lessened. Now it is not the individual fossil that is important but the over all pattern.

However, over time it is not just the scale of the fossil evidence that has changed. The genetics have gradually taken over from the flamboyant and often outrageous personalities of the old-fashioned fossil-hunters. And as the subject has become more scientific it has also run the risk of growing more remote and losing some of its romantic allure and interest to the broader public. We can all empathize with Louis and Mary Leakey and the excitement they must have felt when after thirty years of fruitless toil they at last found an early African fossil human. But can we empathize so easily with the white-coated laboratory scientist who is sequencing a few of the thousands of genes that go to make up the human species, and who is just one of a vast army of similar people?

Gradually the subject of human evolution is being taken away from the glare of publicity and into the recesses of the laboratory. The paradox is that as the subject has become increasingly scientific and so yielded more and more information – for far more about our past is known than ever before, and we can be more confident about that knowledge – fewer and fewer people will have access to the story of our history as a species.

That is why, more than ever, it is vital to make all these findings available. It is the responsibility of the palaeoanthropologists who seek to find out about our past, to bring this story out of dry scientific journals to a wider audience, on television and in books such as this splendid one by Rod Caird. The fossils may belong to the museums, but the story that they tell belongs to everyone. The ideas and the implications that they lead to are part of our heritage.

That is why television programs and clearly written, up-to-date and wide-ranging books are so important, for this is a rapidly changing field. In this book you will find first-hand observations by many of the experts who are working on the many facets of human evolution.

What exactly is the story of the evolution of the most fantastic species ever to roam the planet? It tells us about the origins of being human, and where the species *Homo sapiens* comes from. It tells us of the blind alleys up which the hominids went in evolution, and also the extent to which distinguished palaeoanthropologists followed in pursuit of our ancestors. It also tells us some important things about the way we are today, arising out of our evolution. Perhaps the most important message is that the process of becoming human, of evolving from ape to human, is not that special. The end-product may be – humans are unique and unlike anything else there has ever been in evolution – but the mechanisms of evolution that produced us, the day-to-day and generation-to-generation attrition of natural selection, and the gradual change of genes and species, underlie the evolution of all species, be they protozoa or people.

Thus we discover that this blind and impersonal process produced humans not in a lightning flash, not in a sudden instant of creation, but as the result of accumulation. The origin of humans is not something that can be pinpointed at five million years, or one million years, or 100,000 years in the past, but, rather, occurs continuously over time. Our origin is the whole pattern of evolution, although there are key events that we must discover and identify. The things that make us human are acquired as a complex mosaic – we became upright four million years ago; we began to make tools two million years ago; we began to live all over the earth less than one million years ago; and possibly we only acquired language in the last 100,000 years or so.

Each of these factors is an essential part of the process of becoming human. What makes human evolution such an endlessly fascinating story is trying to visualize the stages, imagining what sort of a creature could walk upright but not talk, make tools but not use fire, survive the rigors of the Ice Age but know nothing of agriculture and a settled way of life.

We learn also that age is not everything. The great temptation for palaeontologists has always been to seek the oldest fossils, to look for still older forms of life, fired by an ambition to show the human lineage stretching further and further back in time. Certainly this was very important in the context of the early days of Darwinism, when many people believed in

the literal and extremely short time-scale of the Book of Genesis. The longer you could prove humans had been on earth, the stronger the support for evolution.

But since evolution as a theory has come of age and is recognized as just a part of the day-to-day fabric of biology, age is perhaps not quite so important. Now it matters less to know *when* something occurred than to ask *why* it occurred. Sometimes this type of question can lead to new perspectives, and one of these is that how young something is can be even more significant than how old it is.

Many scientists today would say that our own species, in its current form, has been around for "only" about 100,000 years. I say "only" 100,000 years in the blasé way that palaeoanthropologists do. It is of course a very considerable period of time – fifty times longer than the period since the Roman empire, for example – but it is also very short in the context of the six or seven million years since the evolutionary separation of the ancestors of chimpanzees and humans. What this means is that all of us have a common ancestor who lived recently, and we are therefore all closely related to each other.

A satellite photograph of the earth, showing Africa, taken in 1985 by the satellite Meteosat.

13

The pioneering work of Jane Goodall, seen here at the Gombe Stream Research Center in Tanzania in 1972, has shown that chimpanzees are close to humans in their behavior as well as their genes.

The things that unite the human race as a species are far greater than the trivial differences that may have developed over only the last few thousand years. We are a young species, albeit the end-product of an older lineage and a vastly older evolutionary process.

Seven million years between us and chimpanzees seems like a long time, certainly much longer than any of us can realistically comprehend. And yet, when we look at that distance in terms not of time, but of the genes that make us what we are, we discover something else which will come as a surprise.

The differences between us and chimpanzees are very small. Not that many genetic changes are required to make an ape into a human. What this tells us is that for all the complexity, for all the enormous achievements of our species, for all the thousands and thousands of genes that go to make up an individual, changing just a few critical ones can have far-reaching consequences. The lesson of evolution might be that very small causes can have major consequences, for while the genetic causes of the differences between humans and chimpanzees may be almost trivial, the consequences for the planet cannot be underestimated.

Of course it takes more than a gene or two to make a new species. It is perhaps not the study of the genetics of chimpanzees that has been most exciting in recent years but the study of chimpanzee behaviour. Through the work of people like Jane Goodall we now know that chimpanzees live complicated lives, almost political in their machinations, that they are capable of compassion, of Machiavellian deception, of immense tenderness and immense violence and cruelty, and that they can and do kill each other. These are all signs that it is not just the genes that bear testimony to our kinship with the apes, but the very roots of our own behavior.

While the geneticists and the primatologists have been exploring this close relationship, the fossil-hunters have not been inactive. Even though we now know that the distance between the living apes and ourselves may not be as great as we once thought, none the less, the space in between them is not empty. Between us and our humanity and the apes and their animal natures, lie the extinct species of "ape man," the ones that did not make it.

Human evolution used to be thought of as a ladder of progress from something bestial and primitive to something advanced, fine and civilized. For a number of reasons we now know that this is not the case. For a start, this is simply not the way evolution works. Evolution does not go in any particular direction, it does not lead inevitably to human beings; instead, it consists of endless trials, endless and often repeated experiments with ways of living, and the diversity of life we see today is the result of all these experiments in every conceivable direction.

Humans are just part of that diversity. What the fossils have shown us is that in our ancestry lie numerous experiments in different ways of living as an "ape man". What you will discover in the pages of this book is not a series of steps in a ladder of progress leading towards us human beings, but instead, fascinating creatures such as the Neandertals or the robust australopithecines which lived for thousands if not millions of years, well adapted to their environment, and not under any great compulsion or urge to "better themselves" and become humans. Whereas in the past we were able to use the ladder as a metaphor for evolution, now we think in terms of a bush with many branches and twigs, and ourselves as just the one surviving twig of a heavily pruned bush.

The story of human evolution that Rod Caird has told both in film and in words is therefore not necessarily what you would expect. There are many new finds, but there is much more that is new too. The modern study of human evolution is much fuller, much more concerned with the whole picture. We have the anatomy of the bones, but we have other evidence as well. The palaeoanthropologists now recognize what ordinary people have always known – that there is more to being a human than just how the bones are put together.

Ape Man

Australopithecus robustus *was, in its own terms, a successful species which survived for around 1 million years in eastern and southern Africa – probably alongside other hominid species. Its eventual extinction was the result not of inadequacy but of the coming of new species, still better adapted to changing circumstances.*

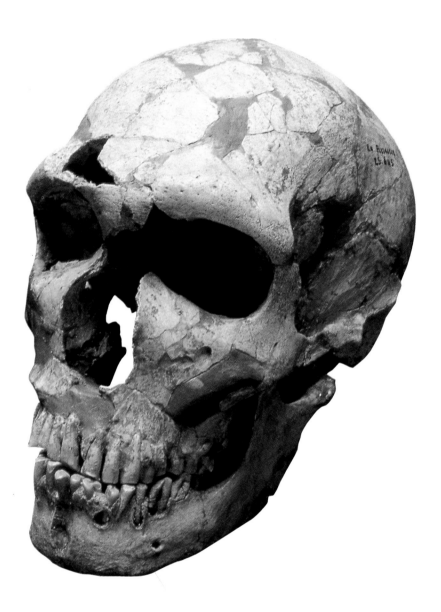

This Neandertal fossil skull was found at La Ferrassie in France in 1909. The Neandertals thrived in Europe for about 100,000 years, finally dying out 35,000 years ago.

And so in these pages you will find out about the evolution of language and tho⌐ art may help us survive, why being intelligent and having a large brain m⌐ tion of better mothers. At heart human evolution is about how pe⌐ they do things, how they survive and thrive in their social lives. A⌐ we look through and beyond the fossils to see the bigger story

When we have told this story we might well ponder abo⌐ sent, and of course we would also ask how the past can guia⌐ pointers forward? Perhaps it is just as well that the answe⌐

know." This is not just ignorance; it is one of the lessons of evolution, for the way a species evolves – and humans are no exception – is the result of the interplay of the laws of evolution with a myriad of chance events.

We can look back and try to make sense of the past, but we cannot look forward with certainty. As a palaeoanthropologist I would say we should be satisfied with this – it is hard enough to reconstruct the past without trying to predict the future.

ROBERT FOLEY
CAMBRIDGE
FEBRUARY 1994

1 *Evolving*

We humans tend to think of ourselves as being completely different and superior to every other creature, living or extinct. We imagine evolution to be a process leading in a fairly straight line to our current state of intelligence, power and control of the environment. It is as though we are the purpose of all evolution.

The truth is very different.

Historically, our story is the probable course of events from the time, seven million years or so ago, at which the ancestral line of present-day humans split off from the ancestors of our closest living relatives, the chimpanzees. Conceptually, the best framework for the tale is an attempt to explain what a modern human is, what the special characteristics are which distinguish us from other animals with which in a number of respects we have a great deal in common.

Evolution has given modern humans two overwhelmingly obvious features which set us apart from other apes and monkeys: we walk upright on two legs, and we have a highly developed brain. Both upright walking and the development of the human brain are inextricably linked to other distinguishing characteristics – the use of language, science and technology. An understanding of how and why these key human features came into existence might provide an answer to the central, puzzling question: why are there human beings at all?

The study of the past involves investigating what has been left behind. There are two main sources of evidence for the story of evolution: material remains in the form of fossil bones and discarded objects like tools; and less tangible relics in the form of the chemical inheritance our ancestors have passed on to our bodies. Modern evolutionary science – and there have been huge changes in this field in the past twenty-five years – draws on both these sources to put together a picture of our history.

Genetically, we are almost identical to chimpanzees. There is less than a 2 percent difference between the DNA of a present-day human and that of a present-day chimpanzee. Indeed, we are closer to chimpanzees than they are to the next nearest relative of us both, gorillas. We and the chimpanzees had a common ancestor, somewhere around five to seven million years ago, and the common ancestors of all three, humans, chimpanzees and gorillas, probably lived about eight million years ago.

In 1959 in the Olduvai Gorge in Tanzania Mary Leakey found a remarkable skull, Zinjanthropus – nicknamed "Nutcracker Man" for its huge jaws and teeth, probably an adaptation to a tough, mainly vegetable diet. It dates to around 1.75 million years ago and was the first robust australopithecine to be found in eastern Africa.

21

The family resemblance between gorillas (above), chimpanzees, and humans is clear enough to an eye willing to look for similarities rather than differences.

Humans, chimpanzees and gorillas share an approximate common likeness. There is no escaping the common-sense conclusion of a relationship between the three, just as there is no escaping the obvious relationship between, say, leopards, tigers and cheetahs. Some kind of family connection between humans and apes was assumed long before the genetic evidence for it was discovered. But it was taken for granted that the human line split away from all the apes at some uncertain point in the very distant past.

The revolutionary new information provided by the geneticists is that humans, chimps and gorillas are a closely related cluster, separated by many millions of years of evolution from other, far more distant cousins such as the orang-utans. Today, the idea of the common ancestor of humans and chimps is well established. DNA and molecular studies since the 1960s have confirmed that the visible similarities between humans and chimps are no freak accident. We and they belong to the same family tree.

There is no fossil evidence of the common ancestor, and until such fossils are found – if they ever are – there can only be speculation about what they looked like. But there is no doubt that, in some ape-like form, the common ancestor did exist.

The span of time since the split with the chimpanzees, in the context of evolution as a whole, is very short. A look at the whole history of life on earth gives a brisk sense of context to the human story.

The dinosaurs died out about sixty-five million years ago, and by then they had been around for 160 million years. To find the beginning of life in any form, the "primeval soup" from which living organisms grew, you have to go back some three to four thousand million years, a length of time which it is effectively impossible to comprehend.

Even taking the extinction of the dinosaurs as a starting-point , and the rise thereafter of the mammals, the seven million years which have elapsed since human ancestors separated

Ape Man

There is still no fossil evidence of the common ancestor of today's humans and chimpanzees. But DNA studies prove that it did exist, and it is reasonable to suppose that it looked something like this reconstruction.

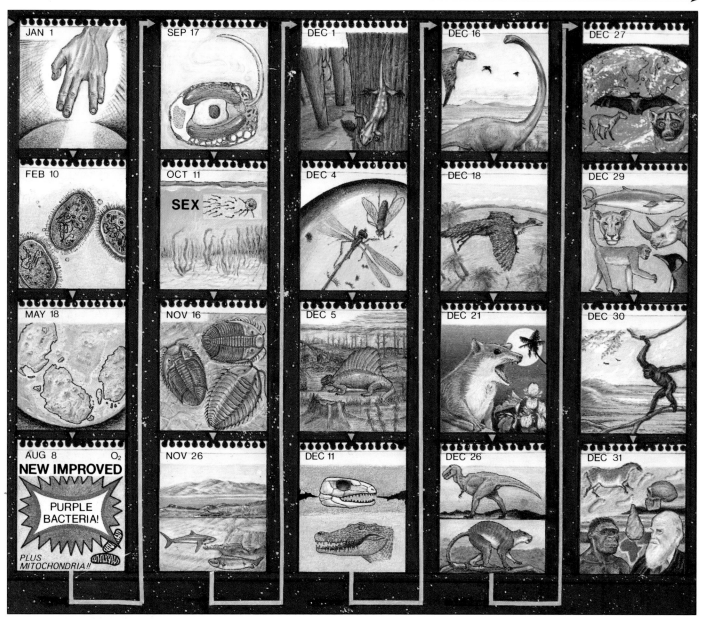

Geological Timescale

The entire span of geological time, beginning with the formation of the earth some 4,500 million years ago, is represented here as a single year. The following events of the geological calendar are shown:

January 1	The earth is formed.
February 10	Single-celled, bacteria-like organisms appear in the sea.
May 18	Most of the earth's land masses are created.
August 8	Purple bacteria appear that can metabolize oxygen.
September 17	Single-celled plants evolve.
October 11	Multicellular plants (seaweeds) are common; sexual reproduction is initiated.
November 16	Complex animals with eyes, legs and brains (eg trilobites) can be seen in shallow seas.
November 26	Jawed fish evolve; plants colonize the land.
December 1	Forests are established; amphibians evolve.
December 4	Winged insects appear.
December 5	Reptiles evolve.
December 11	Largest ever extinction of sea and land animals; crocodiles appear.

December 16	Flying pterosaurs and giant sauropod dinosaurs can be seen.
December 18	Birds evolve.
December 21	Egg-laying mammals and flowering plants appear.
December 26	Last dinosaurs roam the earth; earliest primates appear.
December 27	(before dawn) America and Europe separate; (late afternoon) Lemurs, horses and bats appear.
December 29	(before dawn) Monkeys, penguins, rhinos and true cats evolve; (early afternoon) Whalebone whales appear.
December 30	(early morning) First apes appear; grasslands spread.
December 31	(mid-morning) Origin of hominids; (11.56pm) Modern humans appear.

off from the ancestors of today's chimpanzees represent only the last 10 percent of that time-span.

And, of course, humans did not immediately adopt their present form after the division from the chimpanzees. Most scientists today believe that modern humans – *Homo sapiens* – have been around in more or less their current physical condition for a maximum of about 200,000 years, and perhaps as few as 100,000. In terms of modern behavioral patterns, today's humans could be defined as having a history as short as 50,000 years.

If the mammals took over from the dinosaurs sixty-five million years ago, and it is seven million years since the beginnings of the lines of development from which humans came, a 200,000-year history for modern humans is just a minute blink of the eye. If humans became extinct now, their life-span would seem to future alien palaeoanthropologists like the merest flicker of accidental, momentary development, although these short-term creatures would have left behind an extraordinary wealth of evidence of their lives which would keep those same scientists busy for a very long time indeed.

Archaeology has been defined as the study of other people's rubbish. Imagine the potential difficulty of trying to interpret the durable rubbish of today's humans, and you get some idea of how hard it is to build up an accurate picture of life 100,000 or a million years ago from the random bits of garbage left behind by our ancestors.

Palaeoanthropology is the study of human ancestry. Its intellectual spine is the understanding of how evolution works. Evolution itself is a fact: living things change over time, from generation to generation, in large and small ways. Charles Darwin's contribution, when he published *The Origin of Species* in 1859, was to offer for the first time an explanation of the mechanism of evolution. He made no direct reference to humans, but he came up with the idea of natural selection.

ABOVE: *Charles Darwin, pictured in about 1875, seven years before his death. His grave is in Westminster Abbey in London.*

LEFT: *Nest sites of the dinosaur species* Maiasaura, *dating to about 50 million years ago, were found in Montana in the USA in 1979. They shed new light on the social and family lives of dinosaurs.*

27

Living organisms are constantly producing new features. Many of these are neutral; that is, they are neither beneficial nor harmful. Some of them, however, turn out to work well; the new features fit the environment better than the previous versions and give their owners a reproductive advantage. For the bottom line of evolution is reproduction; anything which increases the chances of an organism producing more offspring will be selected.

Once the new, beneficial feature has been adopted, and has extended to the other members of the species concerned, a change has occurred and evolution has taken place. Natural selection is the motor of evolution.

There are three indispensable requirements in order for natural selection to take place. There has to be variation, a sort of continuous spinning of the genetic wheel producing different features which can be chosen or rejected. There has to be a mechanism for inheritance, so that the new features can be passed on. And there has to be competition, creating a kind of funnel through which only the best-equipped can pass. For if resources were limitless, every variation would succeed.

Evolution is not a thinking process with intentions or long-term goals, whether the perfection of the human being or anything else. It simply happens – inconsistently, at varying speeds, and unpredictably. And this is not a process which happened in the past and has now, as far as humans are concerned, reached some kind of conclusion. Some anthropologists claim that human consciousness and power have reached such a sophisticated level that for all practical purposes evolution has, for us, stopped; we have such control over our circumstances and environment that we can change anything that looks as though it might knock the species off course.

Humans can transform their physical environment. This dam on the Waitaki River on South Island, New Zealand, has changed the landscape to meet the needs of the human inhabitants – at the expense of other animals.

Today, humans can even walk in space. But they do so with a physical form adapted to a life of hunting and gathering in the Ice Age.

Others, however, standing a little further back from the process, observe that our bodies are still adapting to the circumstances in which we now find ourselves. Our physical evolution mostly took place before the great burst of technology in recent times, as a result of which we began to live in huge crowded cities and generally stopped taking serious daily physical exercise. Thus we are adapted not to the way of life we now lead but to a hunting and gathering life-style from 10,000 years ago or more. Who can say what physical changes will be selected by the evolutionary process to make us more suited to today's, or tomorrow's, circumstances?

Natural selection has come up, for example, with the same basic facial shape for all present-day humans: we have relatively flat faces. In our predecessor species, and in the apes, the lower half of the face is more massive and projecting, with much more prominent cheek structures.

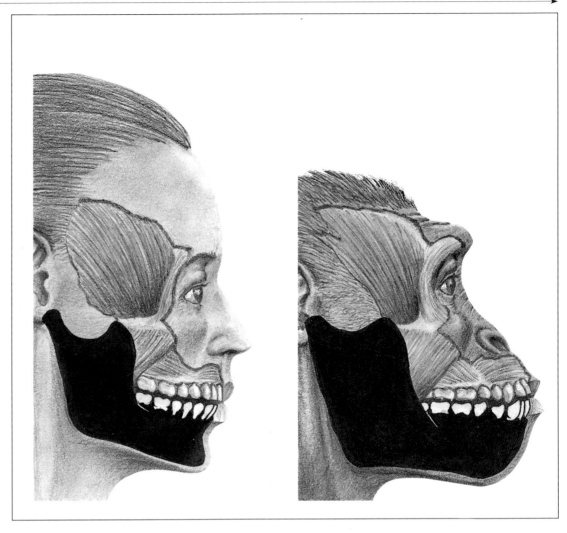

The main changes to the head that have taken place during human evolution are the enlargement of the brain case and the shortening of the face. The reduction of the face is so marked that today wisdom teeth fail to erupt properly in a large proportion of the population.

The modern features have been selected because humans no longer need very powerful jaws or the heavy facial muscles required to chew tough vegetable matter or gristly raw meat. Our faces today reflect the texture of the food we eat – much softer and more pulpy than was the case hundreds of thousands of years ago. By the time we chew it, our food has already been broken down and prepared.

This in turn has meant that we now have a much shorter lower jaw. But we have the same number of teeth as ever, for which there is not really enough space; the wisdom teeth, right at the back, have nowhere to grow. Until the introduction of anaesthetics late in the nineteenth century, allowing their safe removal, impacted wisdom teeth were one of the leading causes of premature death in adults. Infection would set in locally, which would cause further infection and disease elsewhere in the body. For example, people whose ostensible cause of death was pneumonia were often in fact dying of impacted wisdom teeth.

This small example of the wisdom teeth, painfully familiar to all too many people, is highly significant in terms of many aspects of the evolutionary process. A feature is selected (in this case the smaller, more delicate mouth, which works for a specific purpose); it takes no account of any knock-on effect or hidden disadvantage (wisdom teeth); in the modern world, technology intervenes as a short-term solution (dentistry.)

Even if we do intervene in the evolutionary process, even if our relationship with the environment is far from passive, we can never, under any circumstances, opt out of evolution. Nothing can. Evolution will continue whether we humans are around to witness it or not. The only circumstance in which there would be no evolution would be if change ceased altogether – which is unimaginable.

A great deal of the debate about human evolution hinges on "speciation" – change so extreme that a new species comes into existence as a result. Most scientists believe that in the past five million years there have been about a dozen different species in the ancestral line leading to humans – maybe more, maybe less, and, crucially, some of them simultaneous with one another.

If we can find out what brings a new species into life, and what causes its extinction, then much of the mist covering our evolutionary story will lift. The first problem to tackle in this debate is to grasp what a species is.

The original method of classifying living things into groups or species with universally accepted names was invented in 1737 by a Swedish scientist, Carl von Linne (Linnaeus), and his breakthrough completely transformed the study of natural history.

Linnaeus was not concerned with ancestral relationships between species; instead, he took a snapshot of natural history, grouping living organisms into families and giving everything a name. The smallest group, describing individuals so alike that they are identical in practice if not in fine detail, is a species; groups of species belong to a genus; several genera form a sub-family; several sub-families make up a family; and so on, up to the largest umbrella category, the class.

This classification of organisms brought system to the world and made possible the next intellectual leap, Darwin's analysis of how organisms and species change over time. Linnaeus organized the differences between species; Darwin added the processes of inheritance and change, and thus explained how the differences come about.

Every species is unique, in that it is different from all others. There are a number of ways of refining this definition, and not all scientists agree on which is the best one. One common version is that a species is a group of creatures which are distinct from all others in that they can only produce fertile offspring with each other. A less ruthless version would say that a species evolves independently of other species; this definition would not exclude occasional successful interbreeding, but simply makes clear that individuals from one species usually keep clear of individuals from others.

Linnaeus (1707-78) gave systematic names to more than 10,000 animals and plants. Many of the specimens in his collection are now held by the Linnaean Society in London.

A group of males and females of the species Aepyceros melampus, *better known by their common name of impala, a species of antelope.*

Speciation is an extreme form of adaptation. Natural selection produces so many new features, in response perhaps to a major change in a creature's local environment, that the result is a different species. We know that this has happened many times in the human lineage.

The only objective definition of the success of a species is its birth-rate. A species which breeds in large numbers is surviving, always provided its physical environment does not change so suddenly and drastically that every individual is wiped out, and provided also that it does not itself destroy the environment or the food supply on which it depends.

The opposite extreme of adaptation is extinction. Ultimately, most species fail. Chris Stringer, at London's Natural History Museum, gives an explanation of why this point is so important in the context of human evolution: "Extinction is really the rule of evolution rather than continuity, and I don't see why human evolution shouldn't be looked at in the same terms. We shouldn't expect everything we find in the past to be our ancestors."
(Interview, September 1993.)

A fossil which has a rough resemblance to a modern human is not necessarily an ancestor – in fact the chances are that it is not. The number of species which have existed in the past, but which are now extinct, massively outweighs the number alive today.

Elisabeth Vrba of Yale University draws the ideas of speciation and extinction very tightly together: "Speciation and extinction are closely related in terms of cause. Because 90 percent of the events that lead up to either one of these alternatives are the same in both cases. And it is as if near the end the species is poised on a knife edge, with just a blade's-breadth separating the possibilities of extinction versus creation of a new species."
(Interview, August 1993.)

There is a real sense of drama in this notion of success and failure being so close to one another. And the fact that there have been many extinctions in the human ancestral story has the important implication that the principles of evolution are exactly the same for humans as for every other species.

One factor inherent in evolution is the solving of problems. Every evolutionary development comes about as the variations Darwin identified encounter and then solve, or fail to solve, problems in the environment. And evolution does not happen uniformly to entire species; it occurs among local populations in relation to their immediate circumstances – whether it be their food supply, or the climate, or other competing animals in the same area.

"Everything leads on to something else," says Robert Foley of Cambridge University, "so that despite what many people think, evolution is not a perfect little mechanism, perfectly fine-tuning the world; rather, it's very short-sighted. Selection is constantly solving one problem, only to encounter another, and so perhaps our large brains and our technology are just part of that continuous process."
(Interview, October 1993.)

So the story of human evolution is really the story of the evolution of lots of different individual humans in different situations in different parts of the world at different times. And that is why it is so difficult to make grand generalizations from isolated fossils. It is never going to be a simple story.

A skull found in East Africa, for instance, and reliably dated to maybe two and a half million years ago, is only one individual from a species which survived for perhaps two or more million years, and whose population size at any one moment during those two million years is completely unknown.

At this stage we need to clarify some language and ground rules. This is a layman's book, but written as the result of extensive contact with the scientists who wrestle daily with evolutionary problems. Inevitably, there is much technical language in day-to-day use in the scientific community which has very exact meaning but is also largely impenetrable to the outsider. Yet certain terms are impossible to avoid; indeed, being unambiguous, they are extremely helpful.

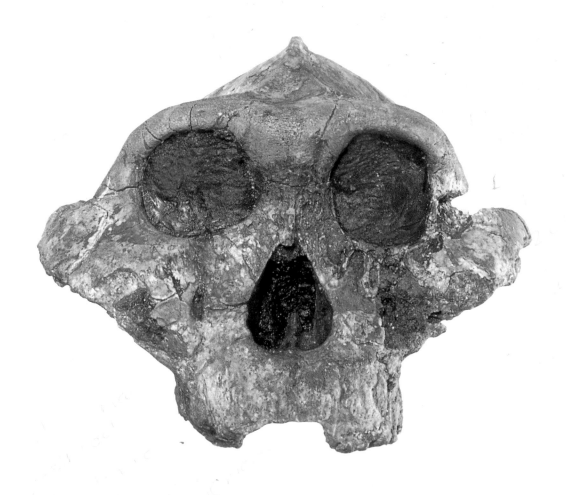

A fossil robust australopithecine skull found in 1969 near Lake Turkana in Kenya by Richard Leakey. Such isolated finds give no clue to the size of the population.

The first of these, which will crop up over and over again in this book, is the word "hominid." This is used to describe all those creatures, preceding and including humans, which have lived on the evolutionary lines between the split with the chimpanzees and modern humans. It is a very valuable term because its use avoids having to attach the label "human" to creatures which may very well be our ancestors but lack some of the characteristics generally accepted as specifically human. In this book "human" is interchangeable with *Homo sapiens* – anatomically modern people, recognizable as such.

Another word scientists use is "hominoid." This describes all apes, hominids and humans collectively. Gibbons (lesser apes), orang-utans, chimpanzees, gorillas (great apes) and hominids are all hominoids. Monkeys, which differ from apes in that they have tails, are themselves divided into Old World and New World monkeys. All monkeys, all apes and all hominids are collectively known as higher primates. And they in turn, together with the lesser primates, which include such creatures as lemurs, belong to the order of primates.

Homo sapiens is thus one species of the genus *Homo* in the family of hominids, in the super-family of hominoids, in the infra-order of anthropoids, in the order of primates, in the class of mammals. That puts humans in their proper biological perspective.

Less scientifically, another linguistic device which needs clarification is the repeated use of words such as "probably" and "in all likelihood." These words are necessary because the science of human evolution is progressing very quickly and inevitably there are many disagreements and uncertainties among the experts involved.

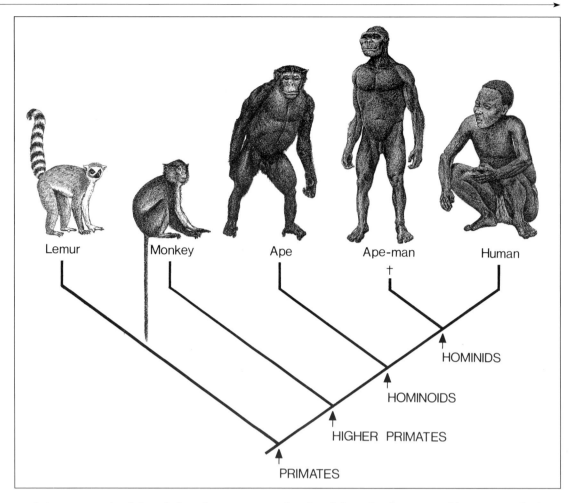

Lemur   Monkey   Ape   Ape-man   Human

HOMINIDS

HOMINOIDS

HIGHER PRIMATES

PRIMATES

Humans, together with the now extinct ape men such as the robust australopithecines, form the group known as hominids. Hominoids are a wider group containing both apes and hominids. A larger group, the higher primates, includes monkeys, and a wider group still, the primates, includes other animals such as lemurs.

The version of human ancestor seen in the film 2001 *was based on an idea of instinctive enthusiasm for personal violence now known to be inaccurate.*

It is commonly claimed that there are very few hard facts in the story of human evolution. This may have been true fifty years ago. But today there is a wealth of generally agreed information, in the fossil evidence, in the more recent study of molecular biology and in the related areas of geology, environment and climate. The common ground among scientists does change, however (the past ten years or so have seen major shifts of emphasis), and in any science which deals with events taking place over millions of years there is bound to be some imprecision.

There have been moments in the past – the scandal of the fake Piltdown skull (see page 68) being the most famous – when science has been misled by an apparent piece of evidence because it happened to fit in with the general assumptions of the time, and such episodes have left a legacy of suspicion which hovers over new discoveries or conclusions.

Equally, there have been new interpretations of genuine evidence which have changed attitudes towards parts of the evolution story. For example, soon after World War II, Raymond Dart found a large accumulation of heavy fossil bones at Makapansgat in South Africa. His interpretation of the find was that they were limb bones from antelopes, used as weapons by early hominids. From this idea came a whole package of theories about the bloodthirsty character of our ancestors, culminating in the brutish image of early man portrayed in the opening sequence of the movie *2001*. Dart's theory was attractive at a time when many people were searching for explanations for human violence: the suggestion was that humans carried an inherited streak of instinctive destructiveness.

But twenty years later Dart's finds were reinterpreted by another South African scientist, Bob Brain. The bones were seen not as carefully selected and stored weapons and clubs, but the natural remnants of wild animals eaten by predators; they had accumulated as a result of leopards and other big cats taking their prey to feeding sites, either in caves or up in trees, from which the detritus would be dropped to the ground. Often, the trees would be growing on the edge of deep shafts; the bones then found themselves as deposits below the ground for later investigators to discover and understand – or misunderstand.

Anyone writing an account of human evolution is therefore bound to use "perhaps," "probably" and "almost certainly" more than is usual. But so many new scientific fields are being drawn into this area that general agreement seems to be increasing. Conclusions no longer rely on one view of a particular fossil; today every new idea is passed across complementary scientific disciplines, starting with traditional archaeology and palaeoanthropology and proceeding right through to the outermost reaches of the analysis of brain function.

Phillip Tobias, Professor of Anatomy at Witwatersrand University in South Africa, and one of the great veterans of the study of human evolution, expands on this: "Today it is unthinkable for a one-man or one-woman show to be on the road in this field. More and more, the advance of our science has become absolutely dependent on many different specialists from many different disciplines . . . It's an interesting change from the old days of the prima donna, the great man who pronounced upon this skull or that skull, to today when a whole variety of different specialists have their input to make into the total analysis'. (Interview, August 1993.)

Leopards often take their kills up into trees to safeguard them. It is easy to visualize how the leftover bones would fall into shafts in the ground, such as the cave at Makapansgat in South Africa.

Louis Leakey was a major star of the fossil-finding world. He would make major claims for his discoveries which attracted massive public attention, but which were sometimes later tempered by less spectacular studies.

So, "probably" does not hide the statement: "This is true if scientist X is right;" it means: "Given that we will never know for certain exactly what happened, this statement is most consistent with all the evidence as it currently stands."

There are two complementary elements in any science. One consists of the rigorous collection and analysis of evidence; the other is the formulation of theories or hypotheses which explain and indeed go beyond the known facts. The first element is safe and sure, and, over time, a picture emerges from the mass of information. On the other hand it is relatively unglamorous and unlikely to lead to great fame for an individual.

The second element inevitably attracts much more attention, and at the same time clearly courts controversy. The history of palaeoanthropology is full of larger-than-life characters who have made assertions about this or that fossil or archaeological find, and have then awaited other studies which either confirm or contradict their insight or opinion. Raymond Dart was a bold proposer of hypotheses, but when Bob Brain overturned his conclusions about violence among early hominids, Dart apparently took it on the chin, as Brain describes: "At that time Raymond Dart made the best interpretations he could and he was deliberately provocative. That is why I believe that Dart was really a great man in science, because he would make statements that would simply demand a response.

"He would make outrageous statements, he would talk about the blood-bespattered archives of humanity – and when you came up with an alternative interpretation, as I did, he was absolutely delighted.

"To begin with he was taken aback, but then he said, 'But you know this is great, this gives a new angle to the whole story.'"

(Interview, August 1993.)

Bob Brain is a mild and gentle man, whose South African excavations over a quarter of a century have been a family affair; his headquarters hut at Swartkrans dig is papered with black-and-white photographs of his family digging with him as well as of palaeontologically illustrious visitors – including Raymond Dart.

Of a very different character is the American scientist Milford Wolpoff, based at the University of Michigan. A noisy bear of a man, Wolpoff has stuck his neck out a very long way over interpretation of the fossil record. His theory flatly denies the modern view that within the past 250,000 years there was a major population move out of Africa which completely replaced the earlier humans already there: in the Wolpoff scenario, all today's populations are descended from groups which have been evolving in distinct regions of the world for four times as long.

The emotional significance of this particular scientific debate derives from the length of time over which different races in the human population have been evolving separately from one another. And Wolpoff's general approach does nothing to defuse the emotional bomb.

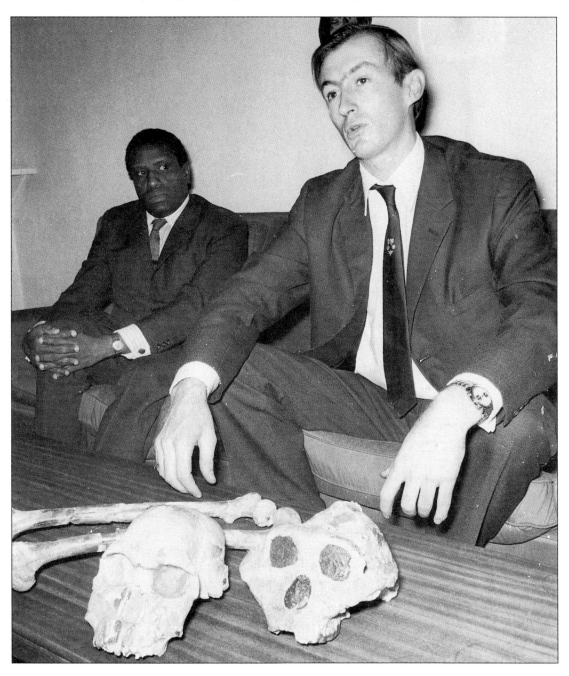

Louis Leakey's son Richard has also made many well publicized individual finds. He is pictured here in 1972 at a press conference in London announcing his discovery of KNM-ER 1470 (on the left), identified as a member of the species Homo habilis.

Phillip Tobias, seen here in the summer of 1993 at the Sterkfontein excavation in South Africa where he has worked for many years, collaborated closely with Louis Leakey. Together with John Napier, they announced the new species Homo habilis *in 1964.*

"Science can only work by refutations. We'll never know what's right, we'll know what's wrong. And opposing hypotheses are particularly important because each will focus on the weak spots of the other hypotheses; they'll give us the place to look. I tell my students that when they get in a scientific argument, they should know the opposition better than the opposition knows itself."
(Interview, August 1993.)

Elisabeth Vrba, who has made a special study of the relationship between major evolutionary changes and the climate, summons the spirit of a modern European philosopher to lend support to her own ideas.

"The Austrian scientist Karl Popper came along, and he said the glory of science is not to amass more and more facts. Our only hope in comprehending nature is to make wild, unsupported suppositions – to be fearless, to leap across the discontinuity of lack of knowledge. To try out totally new ideas, make them as bold as possible, and then look for the evidence to see whether we need to accept them or reject them."
(Interview, August 1993.)

It is an intriguing reflection of the current character of the study of human evolution that Vrba's bold ideas are not primarily to do with the significance of individual fossils but with identifying climatic change as the motor force of evolution.

But the principle remains the same. Evolutionary science is littered with colorful and imaginative individuals whose driving force is the desire to understand and explain not some dry mathematical theorem but the meaning of human life. Mistakes may have been made, and today's state of knowledge may be much closer to an accurate picture of the past: but there remains the common thread that this is the story of ourselves.

As Robert Foley puts it: "The study of human evolution has its own character and its own culture. Palaeontologists in general want things to be older than current evidence would suggest. No one wins prizes for finding the second oldest fossil. There is an innate tendency to seek origins going further and further back. That's reinforced by our own assumptions; we think of ourselves as very different from all other species living today.

"There's an underlying assumption that we should go further and further back to find the last common ancestor, to find the first hominid.

"But ironically we're moving to another stage now; what's striking is that things are much younger than we thought. We no longer think that the first hominid existed fifteen to twenty million years ago, but perhaps seven million years ago. We no longer think that the first member of our own genus, *Homo*, goes back two or three million years; we talk more about *Homo sapiens*, 100,000 years ago, being the first human.

"We're in a very exciting phase, reconsidering what it means to be very ancient, and what it means to have events that are very young, very close to us."
(Interview, October 1993.)

Walking

In 1974 the American palaeoanthropologist Donald Johanson, investigating fossils in Hadar in Ethiopia, found almost half the skeleton of a complete individual. In fact he had been on his way with a colleague to look at the fossil remains of an ancient pig – interesting, but not earth-shattering – when they saw a collection of other, much more exciting bones scattered on the surface of the ground. For millions of years the bones had lain buried where their owner died; erosion had brought them back to the surface.

The creature concerned was small, female, and from the shape of the bones undoubtedly walked upright – so she was a hominid. Johanson called her Lucy, because the Beatles' song "Lucy in the Sky with Diamonds" was playing in the camp-site on the day they brought her back.

Lucy's pelvis and lower limbs displayed all the adaptations necessary for upright walking, or bipedalism. These parts of her skeleton were more like those of a modern human than those of a quadrupedal ape. Her pelvis had partly acquired the bowl shape of human anatomy, as distinct from the elongated appearance of an ape pelvis. And her heel bones had developed the energy-absorbing structure essential for any creature in the habit of putting its entire weight on one foot at a time.

These features make Lucy a hominid; she is a member of the human family. And her lower jaw showed her to be different to any other hominid species found up until that point, so Johanson gave her a new species name – *Australopithecus afarensis*, after the Afar, the region in which she was found.

In palaeontology, the finder has the privilege of giving a new species its name. This does not make him or her immune to the criticism that the new species is nothing of the sort – but it has led to some inventive (and often tongue-twisting) terms.

The word *Australopithecus* was coined by Raymond Dart in the 1920s, and he got into a lot of trouble for it. It means "southern ape," and is a mixture of Greek and Latin. One of the many sticks with which Dart was beaten by the palaeontological establishment of the time, was that it was etymologically incorrect to mix up language like that, and anyway the word was cumbersome and hard to pronounce.

Much of this criticism was code for the establishment's more general dislike and suspicion of Dart and all his work – in fact, there are no formal rules about choice of language for

ABOVE: *Donald Johanson with Lucy – the earliest fossil found of an upright walking hominid.*

OPPOSITE: *In 1978 Mary Leakey found a trail of footprints at Laetoli in Tanzania which had been made 3.5 million years ago. They are still the earliest evidence of upright walking by hominids: the "smoking gun" which shows the time of the origin of bipedalism.*

41

Ape Man

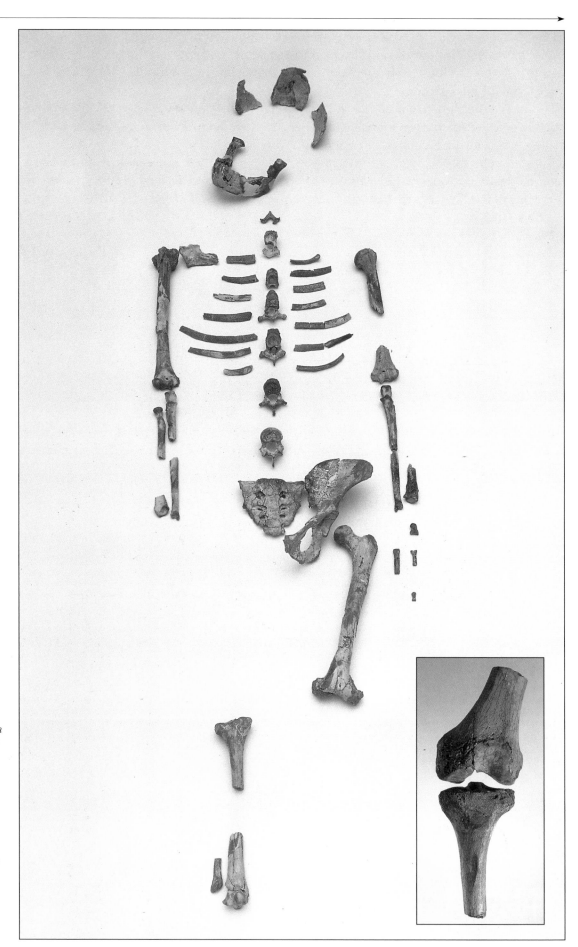

Lucy was small, no more than about 4 ft (1.2 m) tall. Other individuals from the same species, Australopithecus afarensis, vary from about 3 ft 3 in (1 m) in height to about 4 ft 11 in (1.5 m). Their weight when alive ranged from about 66 lb (30 kg) to as much as 154 lb (70 kg). An A. afarensis locking knee joint from the same area (inset) provided a clear clue that its owner walked upright.

species' names. In essence, you can call species anything you like, as long as the name is different to any other. In any event, Dart's tortured word *Australopithecus* has stuck. Johanson added *afarensis* to distinguish the Lucy species from Dart's *A. africanus*, which had been named two decades earlier.

Dating of the rock structure where Lucy was found put her at about three million years old. This meant that she was more ancient than any other hominid previously discovered.

Owen Lovejoy, an anatomist at Kent State University in Ohio, knows the story of Lucy and Johanson well. In 1973, Johanson had brought a fossil knee joint back from Ethiopia. Lovejoy says it was one of the most dramatic events of his life, because until that moment there was simply no evidence that any creature that old had walked upright.

"So I said, 'You know, if you can find fossils this good why don't you go back and find a skeleton?' And the next year he actually did, he found Lucy. I said, 'Well, maybe I should have something to compare it with, so go back and find me a population.' And in fact he did – the next year he came back with mom, pop and the kids."
(Interview, August 1993.)

Donald Johanson tells the dramatic story of the new find in his book *Lucy, the Beginnings of Mankind* (Donald C. Johanson and Maitland E. Edey, Penguin, 1990): "Finding teeth at the bottom of the slope and two other bones halfway up encouraged a closer scrutiny of it. Almost immediately, other fragments were spotted. Looking on the other side of the acacia bush, I had the unnerving experience of picking up, almost side by side, two fibulas – the smaller of the two shin bones in the leg.

"Another Lucy? No, these were both right legs, indicating the presence of more than one individual. Meanwhile, others were shouting over finds of their own, all of them hominid. Fossils seemed to be cascading, almost as from a fountain, down the hillside. A near frenzy seized us as we scrambled madly to pick them up.

The A. afarensis *finds came from Hadar, in the Afar region of Ethiopia, where Johanson continues to excavate. In 1994 he announced the discovery by Yoel Rak of Tel Aviv University of the first relatively complete fossil skull of the same species, dated to 3 million years ago.*

Fossil jaw bones of A. afarensis *have much in common with those of the apes in the size and shape of the teeth and the structure of the face.*

"For a little while I scarcely knew what I was doing. I had never seen anything like it. I had never heard of anything like it. We were like crazy people. Finally the heat got to us, and we settled down."

The new find, in 1975, from what became known as the First Family site, yielded over 200 fossil specimens and helped to build up a fuller picture of the species as a whole. In particular, whereas most of Lucy's skull had been missing, the First Family site provided much more information about the shape of the head, proving, for example, that however advanced *Australopithecus afarensis* may have been in their two-legged gait, their brains were still very small when compared with body weight.

The shape of the head was still relatively ape-like. The face projected forwards, the brain was small and the canine teeth were quite large. Proportionately, the legs were quite short and the arms quite long. *Australopithecus afarensis* were fully bipedal, but they still had a lot in common with their ape-like ancestors.

How long *Australopithecus afarensis* prevailed for is not known. There is a gap in the fossil record backwards from Lucy to the split with the chimpanzee ancestors. But other discoveries, in both Ethiopia and Tanzania, suggest that the species existed for perhaps a million years (far longer than *Homo sapiens* has so far managed) and that it was fairly well spread out geographically through East Africa.

One of the great excitements of the Lucy find was the close dating match with the information provided by molecular research about the common ancestor. The molecular biologists had offered a date of five to seven million years ago for the split between the hominid and chimpanzee lines; here was a bipedal creature, with strong ape-like features as well, dating at three million years ago. The fossil and molecular evidence seemed to be telling the same story.

Lovejoy believes, from a study of Lucy's leg and foot bones, that by the time she was alive bipedalism had been established for as long as two million years. This was no halfway-house species, trying out a new way of getting around with a staggering gait. Lucy and her relatives, according to Lovejoy, were confident and effective on two legs and came from a long line of upright ancestors.

This view is not universally shared. Bernard Wood, of Liverpool University, says that Lucy's limb proportions and skeleton suggest she was neither predominantly tree-living nor fully upright, that she and her relatives "may therefore have fed and moved much like modern baboons, but with more emphasis on bipedalism. Such groups would have spread out to forage on the ground in the day, and then congregated, perhaps in caves or trees, at night." (*Cambridge Encyclopedia of Human Evolution*, Cambridge University Press, 1992.)

The fossil hand bones of A. afarensis *show that members of the species were probably almost as dextrous as modern humans. In the picture ancient fossil bones, recognizable by their colour, have been included with modern bones to indicate the closeness of the fit.*

This view centres on the length of Lucy's arms and her curved hands. Bill Kimbel, who works very closely with Johanson, and is Director of Palaeontology at the Institute of Human Origins in Berkeley, California, says the fossil evidence thus far cannot provide all the answers. There is no doubt that Lucy walked upright, but according to Kimbel, whether she was also able to get around in the trees, either to sleep or to avoid predators, can only really be speculated about.

Lovejoy rebuts this view. He says that the short legs were simply less prone to injury; the long arms were good for dispersing heat. The absence of grasping toes made Lucy unable to climb trees effectively.

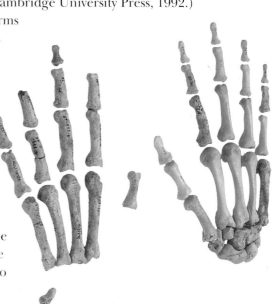

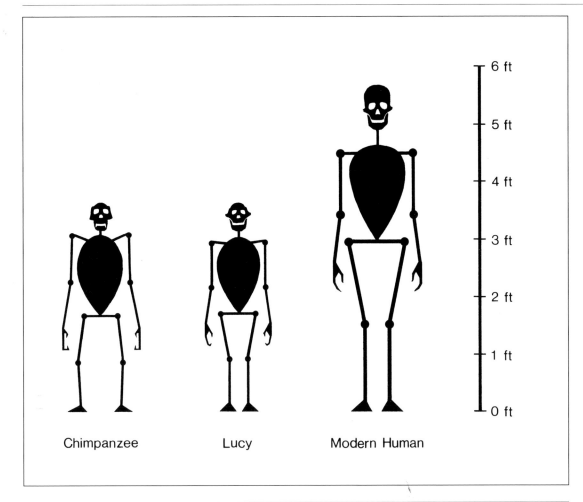

Chimpanzee   Lucy   Modern Human

Lucy has the human-like characteristics of upright stance, arched feet, and relatively long thumbs. But she also has the ape-like features of small size, short legs and long arms, a small brain, somewhat hunched shoulders, and moderately curved fingers.

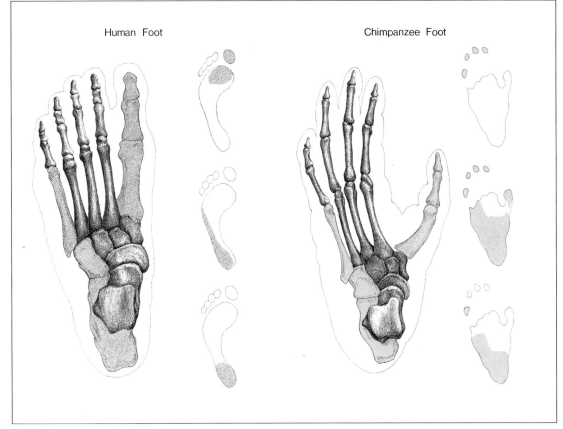

Human Foot   Chimpanzee Foot

The human foot is strengthened (see stippled bones) to withstand the pattern of pressure generated during walking. Footprints show that the pressure begins at the heel and then moves up the side of the foot before transferring to the ball and finally the big toe. The human heel bone and big toe are especially large and robust compared to those of a chimpanzee, which transfers its weight differently.

Lucy and other members of her species lived in an environment very different from today's arid Ethiopian climate. They probably lived alongside rivers where there was abundant wildlife – and may even have been preserved as fossils as a result of drowning, perhaps in flash floods.

There is only so much that the fossils can tell us. They are not in a position to answer back, or to wriggle in embarrassment at some wild misinterpretation of their meaning. They just sit there while the arguments wash around them – after all, even their original owners did not know what species they belonged to.

Common sense and simple arithmetic lead to the conclusion that over that period many hundreds of thousands, probably millions, of Lucy's relatives existed. In an article in the *Cambridge Encyclopedia of Human Evolution* Bernard Wood says that the entire australopithecine fossil collection represents between 0.02 and 0.00002 percent of the estimated living population of those species.

Lucy herself was not necessarily any more representative of the entire species of *Australopithecus afarensis* than any one individual today is of *Homo sapiens*. In the years after Lucy's naming, controversy raged over whether she and the First Family remains represented one species or several, centering on the debate about how much difference there needs to be between individual specimens before they constitute separate species.

The alien palaeoanthropologist of the future may well look at the partial skeletons of an Eskimo, an East African and a Dane – very different in size, limb proportions, and facial features – and puzzle over whether they could possibly all be *Homo sapiens*. But there is no doubt whatever that they are; whatever definition of a species is chosen, it is impossible to find any biological barrier between any two modern humans. So today's scientists have to take account of likely wide variations in appearance within a single species, although it is also true that the variations will probably be less if the species lived within one small area, without the big differences in climate and general environment which can wreak superficial physical changes.

The fact that the fossil record is still relatively small explains why there is such excitement, drama and pride when anything is found which adds to existing knowledge or changes previous ideas. Most remains are not preserved for future investigators to discover. Instead, they have been destroyed by predators, the weather, the movement of the earth into which they sink, or by all three. Humans did not start carefully burying, and thus preserving, their dead until very recently (about 100,000 years ago); before that, the remains of our ancestors took their chances on open ground or in caves.

Anyone who has seen natural history television programs showing predators hunting and eating their prey will know how little is left of the victim – especially when the predators consist not only of the killer lion or cheetah, for example, but also the vultures and hyenas which arrive in their wake to feed off the carcass of the unlucky gazelle.

Andrew Hill, a highly respected scientist with many years of experience of excavating in East Africa, puts it like this: "When animals die, they tend to rot and disappear and get fragmented and thrown away. Very, very few of them become buried, turn into fossils and are then discovered by people like myself. So it's really very surprising that we know anything about our own ancestry. We have no right to expect that."
(Interview, August 1993.)

It is likely that Lucy was not the victim of a predator, but died of some other cause such as accidental drowning; thus her skeleton was not destroyed or scattered, but was preserved for Johanson and his colleagues to find about three million years after her death.

And the so-called First Family fossils provided relics of thirteen individuals which could not all have been killed simultaneously by predators; nor is there any sign that they died in some sort of prehistoric battle. More likely, they were drowned, perhaps as the result of being caught in a flash flood, a fate still well enough known and feared today by nomadic people.

To become a findable fossil, a bone has to sink into the ground, or be otherwise covered over, quite soon after its owner's death; it remains buried until erosion brings it back to the surface or underground excavation discovers it still embedded in the earth. And the movement of the various layers of material under the surface of the earth will further disperse or

Most bones are destroyed before they can become fossils, and even then most fossils will never be found.
1 *A hominid is ambushed and killed.*
2 *The carcase is dispersed by scavengers, and the river current drifts the head downstream.*
3 *After annual flooding, some of the bones are buried under layers of silt. Others are broken up by trampling or the weather. Once a bone is completely buried, its protein material is replaced by minerals from the surrounding water – fossilization has begun.*
4 *Millions of years later, the sediments in which the fossil skull is embedded have been thrust upwards by geological forces.*
5 *The sediments are exposed to erosion, especially by wind and rain, and eventually the skull is washed out – to be found, perhaps, by a sharp-eyed fossil hunter. If not, it will be buried again, by fresh slope wash, within a year or two.*

Raymond Dart holding the fossil skull of the Taung baby, which he found in 1924.

shatter the bone. So it is easy to see why a fossil find, especially of something relatively complete, causes such excitement.

Most fossils today are studied in the form of accurate casts; the originals are just too fragile and precious. But in a wooden box, locked in Phillip Tobias's safe in the Anatomy Department at Witwatersrand University, is the original Taung baby, a tiny skull found in South Africa by Raymond Dart in 1924, an australopithecine discovered long before Lucy but not accepted as such until the 1950s. When Tobias gets the original fossil out of its box and holds it, as he must have done many hundreds of times, he says: "It still is the most thrilling experience. The hair on the back of my neck stands on end still every time I handle this little child. It's a wonderful feeling to be able to hold such a little child in one's hands, and to realize that this creature, two and a quarter million years old, was the find that triggered the whole subsequent history of human evolutionary studies in Africa. There was nothing before this and there was everything after this.

"[The fossils] are world crown jewels, and it's inevitable one's going to have a sense of awe, and the great thing one must never lose is one's sense of wonderment."
(Interview, August 1993.)

Bob Brain who, like Phillip Tobias, has worked all his life in South Africa, devoting much of his time to his single-dig site at Swartkrans, is similarly eloquent: "Palaeoanthropology is a curious subject, in that the actual fossil remains tend to be rather scarce. There are far more palaeoanthropologists than there are fossils to go round, and so there will always be lots of ideas. And it's an emotional subject, because one is dealing with one's own ancestry, with people.

"When you've spent a lot of time and a lot of effort digging something out of the ground or even fortuitously finding it, then it's natural to feel this is almost like one of one's children, and I think there's a tendency to make it the most significant, the oldest fossil that ever was.

"I think [the scientists] should always be great enough to be able to disregard [their theories] when new evidence comes along to provide a different interpretation. And that's going to happen regularly in a science like this."

(Interview, August 1993.)

There is no denying the significance of Johanson's discovery of Lucy and the other australopithecines in Ethiopia. As we've seen, these fossils established that hominids were walking upright at least three million years ago. On the other hand, their bipedalism and their slightly reduced canine teeth were the only features which put them on the hominid line. It is possible that they could have used very simple tools, rather as chimpanzees do today: sticks to poke into termite mounds, or rocks to open fruit. But there is certainly no evidence that they shaped stone tools, and such sophistications as the use of fire still lay well in the future. Their brains were the size of an ape's, and details of their daily lives or social organization are unknown.

In its own terms, *Australopithecus afarensis* was a successful species. Some people would call them primitive – but the word carries a value judgment. It suggests that they had not yet progressed to where they "ought" to be, that they were not fully developed. This way of thinking relates to the notion that all the early hominid species can be placed on a ladder, leading to the triumphal top rung where *Homo sapiens* sits proudly looking down on his or her imperfect predecessors.

Evolution is not like that at all. *Australopithecus afarensis* was fully developed. So is *Homo sapiens*. Both deserve to be studied in their own right, as separate adaptations to the environment of the day, as well as species which are undoubtedly related to one another, possibly even in an ancestral line.

Robert Foley provides an interesting analogy. He says that too many people study human evolution as though they were reading a detective novel by starting with the final chapter. Everything they read thereafter is retrospective; they see it in terms of what happens at the end of the book. But in a good novel, the punchline is irrelevant to the overall development of the characters in the story. Why a species such as Lucy's thrived and then finally failed deserves to be studied in its own right, not as a step on the road to becoming human. In any event, we are not yet at the end of the story.

Chimpanzees can regularly be observed using simple tools to get at their food.

Ape Man

The trail of footprints discovered at Laetoli by Mary Leakey was about 150 ft (70 m) long. The prints show a well developed arch to the foot and no divergence of the big toe. They are of two adults, with possibly a third set belonging to a child who walked in the footsteps of one of the adults. The prints on the far right are those of a hipparion, an extinct three-toed horse.

There is no such thing in evolution as an intermediate species. As evolution does not "know" what is going to happen next, and is not even "thinking" about it, then by definition each species is exactly as it should be at any given time. It is not "going" anywhere.

Four years after Donald Johanson found Lucy, another of the giants in this field made an equally extraordinary discovery. Mary Leakey was excavating at Laetoli in Tanzania when, on the morning of 2 August 1978, she uncovered some footprints preserved in volcanic ash. The moment is described by John Reader in his book *Missing Links* (Penguin, 1990): "At 10.45 Mary Leakey straightened up abruptly. She lit a cigar, leant forward again, scrutinized the excavation before her and announced: 'Now, this really is something to put on the mantelpiece.' "

The prints had been preserved by an unusual set of circumstances. Near Laetoli is a volcano called Sadiman, whose ash sets like concrete if it gets wet. It seems that an eruption of the volcano three and a half million years ago, spreading ash over the Serengeti plain, was followed by a shower of rain. The marks of some of the rain-drops can still be seen. Twenty different species of animal, including hominids, then walked over the damp ash, and their prints, later covered over by more ash and silt, set rock-hard.

The first animal prints were seen by Andrew Hill in September 1976. Other discoveries were made in the same area, culminating in Mary Leakey's find of a whole trail of hominid footprints, showing that two individuals had walked that way, possibly accompanied by a third, smaller figure, walking partly in the footprints left by the others. The volcanic ash in which the prints were set was known to be more than three and a half million years old – and the shape of the prints was for all practical purposes that of a modern human foot. The indentations, the shape and distribution of the toes, the relationship between the marks left by the heel and ball of the foot all showed that the individuals who walked that way had acquired the fully upright, two-legged gait used by humans today.

Owen Lovejoy describes the Laetoli footprints as the "smoking gun" evidence of bipedalism. "With the footprints we have really the ultimate form of evidence of upright walking. The footprints show that the longitudinal arch was present, that the great toe had moved in line with the other digits and was no longer opposable, and that the animal had a very definitive heel strike similar to that which we see in modern man.

"Together this indicates that we're dealing with an animal that had been upright walking for a very long time prior to the making of the footprints."
(Interview, August 1993.)

The arch of the foot is crucial. It is an adaptation which gives a spring effect, spreading the force of the foot hitting the ground. The human foot lands on its heel and rises up again on its toe; the arch, in between, acts as a spring. This is unique to humans, and the evidence of the arch in the Laetoli footprints clearly distinguishes them from any other animal.

The walk may even have been more elegant than ours. The small brain size of *Australopithecus afarensis* meant that mothers would probably have had narrower birth canals, and hence narrower hips than modern human women. This would probably have given them a very smooth and even gait.

The footprints put bipedalism even earlier than the fossil evidence of Lucy had indicated: by at least three and a half million years ago, the transformation from four-legged, tree-living ape to two-legged hominid was complete. And that of course means that the process of walking upright had begun long before that date; the detailed physical changes to feet, legs, pelvis and spine must have taken many thousands of years to complete.

The Laetoli footprints are not on anyone's mantelpiece. They are still in the ground in Tanzania, where Mary Leakey found them. Ironically, there is currently controversy about their state of preservation. After they were first discovered, they were carefully covered again to protect them from the weather and from the wear and tear of visitors. But many people in the academic community are extremely concerned that one consequence of Tanzania's current political and economic difficulties may be that the site is not being cared for as well as it might be.

There had been a plan, cancelled at very short notice, that the footprints should be uncovered again during an academic conference in the summer of 1993. The Tanzanian authorities will not at present allow anyone to visit or film at the site. There does appear to be a risk that the prints, preserved by nature for over three million years, may be lost by lack of attention in a mere couple of decades. Scientists are understandably disturbed at the prospect that the world's oldest relic of hominid life may disappear; non-scientists ought to be equally worried.

In any event, the photographs of the originals, and casts which have been made of them, send shivers down the spine. Because they are imprints made by living flesh, they have a special power. It is easy to imagine the people who walked that way, irrespective of what they may have looked like exactly. True to form, however, Mary Leakey is somewhat relaxed about her discovery today: "I suppose by the time I found them I was a bit blasé, I'd found other things before. [But] they establish, I think without any question, that as long ago as three and a half million years our ancestors were fully upright with a bipedal stride like our own and feet that are almost indistinguishable from ours today."
(Interview, August 1993.)

As the Laetoli footprints were probably made by hominids from the same species as Lucy, and other fossils similar to Lucy were found in the same area, scientists believe that *Australopithecus afarensis* survived for at least a million years, and that they occupied a substantial area of Africa.

The obvious question is: why did Lucy's species walk upright? What evolutionary problem was being solved by it being better for *Australopithecus afarensis* or its predecessor species to stand upright on two feet than scurry about on four? These are by no means simple questions, and there are very nearly as many answers as people to ask.

You only have to watch chimpanzees in a zoo to realize that their means of getting around is very effective. They are fast and highly flexible; you can easily imagine that they are well equipped to get out of the way of predators by four-legged running on the flat, or climbing in the trees, or a combination of the two. They are sufficiently well balanced to gather food with one hand while moving around, or to stand upright for brief periods, using both hands if they need to reach up for something.

They use simple tools to get their food, to break open nuts or to poke into termite hills, and they have no obvious problems doing what they need to do to survive. There is no evident pressure on them to do things differently. So why, if we share a common ancestor, did we become bipedal while the chimps did not?

The question is based on a false premise. It makes the misleading assumption that the common ancestor of humans and chimps was like today's chimpanzee. On the contrary, modern chimps have evolved from that point to their current condition, just as we have; they must have been changing, over about the past seven million years, as hominids have changed. In between the common ancestor and today's chimps were presumably a variety of species with a variety of adaptations.

In evolutionary terms, today's chimps are less of a success story than hominids, but they are by no means a failure. Until very recently, they were common throughout West and Central Africa – a span of about 3,500 miles. Their means of locomotion is extremely effective in their wooded or forested environment. Moreover, there are two chimp species – the common and the pygmy. The one thing they lack is the great behavioral flexibility of their cousins, the humans.

The common "ape" ancestor was different to both chimps and hominids. But it must have got around in an ape-like way – on all-fours, in the trees and on the flat: so what was the problem being solved when the first hominids began walking upright?

Being two-legged is not all good news. In fact, from some perspectives, it seems thoroughly negative. Bipedalism is not necessarily a wonderful way of getting out of the way of sabre-toothed tigers. It does not make you faster or more agile. It would not, therefore, make you any better at catching things for lunch, or, indeed, at avoiding becoming lunch.

It puts appalling strains on elements of the skeletal frame, most particularly the backbone. Any family doctor, anywhere in the world, will tell you how much pain, suffering (and lost working hours) result from weaknesses of the back caused by the strain of maintaining (or failing to maintain) a relaxed upright posture.

It requires changes in the shape of the pelvis which make childbirth more difficult and dangerous for humans than for any other known species.

It makes human babies more vulnerable – and for longer – than the offspring of any other species. Human babies are not safe on their own two feet until they are at least two years old; chimp babies can get about very effectively within months of birth.

Two legs are very prone, even in adults, to accidental damage. The knee is a fragile joint, and the body weight is heavily concentrated on each foot as it lands on the ground. If your foot lands awkwardly on uneven ground, even when you are walking quite slowly, you can completely incapacitate yourself. This is a bad enough problem today, but it does not take much imagination to see what would happen to a lame hominid three million years ago in the open African savannah. Whatever the reasons for being upright, therefore, they must have been good ones, for the price has been very high.

The storybooks on evolution recount how the forests of Africa retreated, giving way to savannah; our ancestors had to be able to stand upright in order to watch out for predators over the long grass. Yet Lucy was very short, standing not much more than a metre tall.

The area of Ethiopia where Lucy lived is today very dry and hot. There is one river running through it, but any distance away from its bank the atmosphere is harsh. In Lucy's day, according to Bill Kimbel it was a cooler, wetter place:

Conditions in the Okavango Delta in Botswana today may quite closely resemble those found further north and east in the days of Australopithecus afarensis *3 million years ago, with wooded, relatively wet areas mixed with expanses of savannah.*

"There was a large lake that fluctuated in size over time, and river systems coming from the highlands, feeding the lake. The early hominids, together with the very rich local wildlife, lived along the water systems that flowed in and out of the area.

Ape Man

The map on the left shows Africa about 20 million years ago, when this continent was warmer and wetter than it is today. Tropical forest extends in a broad belt across the continent, with grasslands only at the edges. Ape evolution is in full swing. By 6 or 7 million years ago (right), much cooler and drier weather has made the forest contract and the dry savannah has advanced from north and south. A huge rift valley has opened up along the eastern side of the continent. This eastern region now contains savannah, lakes, and forest fragments. Hominids are established as a lineage adapted to the new conditions.

"[Lucy] was probably a generalist in terms of dietary choice, an omnivore. Although she was a biped, and successful at that, bipedalism is not a particularly apt way to get about, if avoiding predators is your goal.

"And no doubt Lucy and other members of her species fell prey to the carnivores that inhabited the community in which they lived . . .

"The question as to why Lucy stood up is a very difficult one. There are a number of competing ideas in the community as to what the advantage might have been, some of which sound credible and some of which sound incredible."
(Interview, August 1993.)

Elisabeth Vrba has identified two periods of huge overall climatic change: "One of these is the time between seven and four and a half million years ago, and the second is between three and two million years ago. Both were times of dramatic cooling, major climatic changes, and of massive evolutionary changes right around the globe, both in the ocean and on land.

"And it so happens that these are the precise two time periods that have witnessed the most dramatic advances in human evolution. First, bipedalism arose between seven and about four and a half million years ago, and second, the advent of the expanded brain of the genus *Homo* and of stone tools, nearly two and a half million years ago."
(Interview, August 1993.)

Vrba says that the evolutionary process over the last sixty million years has taken place against a background of steady cooling of the globe, and that within that cooling there have been frequent, dramatic fluctuations. A major one of these occurred between seven and five million years ago.

"The evidence is accumulating from all continents and all oceans at this moment. The oceans became about five to ten degrees cooler, the average temperature of the globe became considerably cooler. On a land mass like Africa, one of the major effects was that the

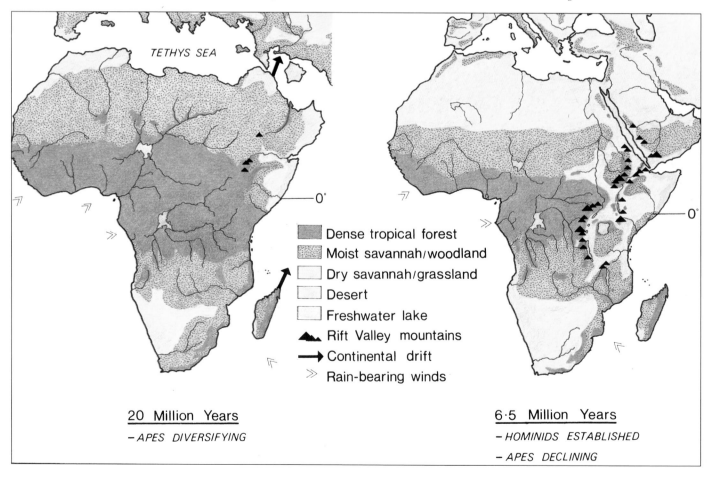

TETHYS SEA

0°

- Dense tropical forest
- Moist savannah/woodland
- Dry savannah/grassland
- Desert
- Freshwater lake
- Rift Valley mountains
- Continental drift
- Rain-bearing winds

0°

20 Million Years
– APES DIVERSIFYING

6·5 Million Years
– HOMINIDS ESTABLISHED
– APES DECLINING

An area of mountain rain forest near Mount Kenya.

warmer, wetter, more forested areas, habitats, were shrinking wholesale. Climates, in most places, became much more seasonal. Food and water would suddenly not be there, all the year round.

"Tree cover and forest cover in Africa shrunk very dramatically. Any creatures that were dependent on sheltering in the trees and on fruits and leaves from the trees as food would be dramatically affected by this.

"Suddenly these early creatures, the ancestors of which used to live in the trees, in the high-canopy forests of Africa, found themselves in much more open savannah and woodland situations. And they had to adapt or die. And I believe that bipedalism may be a direct response to this opening up of the vegetation."

(Interview, August 1993.)

This argument depends not so much on the idea that getting around on two legs is better or worse than getting around on four, as that getting around on four is specifically linked to a successful life in and around the trees. When the trees retreat, or grow further apart, the specific characteristics designed for tree life (long and powerful arms, fingers and hands shaped for gripping branches) become redundant and are replaced by a different shape more suited to the new conditions.

Before the big climatic change identified by Vrba, forests spread right across Africa; now they are concentrated in the center and west of the continent, where the four-legged chimpanzees and gorillas live. As the forests retreated, Lucy and the other members of her species adapted to a two-legged life in the open.

Vrba says that she was taught in the Darwinian tradition of the inexorable, relentless process of evolution. But when she went into the laboratory at the Transvaal Museum in South Africa for the first time, she was struck by the evidence of the absence of change. The fossil record shows that for many millions of years in some cases the physical appearance – the morphology – of many creatures stays pretty much the same.

Typically open savannah in East Africa.

Vrba's new theory is that evolution is in fact a very conservative process, "because organisms and species are complex systems, with many interacting sub-parts. And for these systems to function, they require very particular external circumstances. From this derives the idea that unless those external circumstances change, the living systems will remain conservative."

The pacemaker of evolution would be any external circumstance which affects the habitat of a species. "The most conventional way of understanding that is climate – whether it rains more or less and whether the temperature's higher or lower. But we have to think a little more broadly to understand evolution and include all kinds of tectonic change, by which I mean, for instance, a rift valley opening, as we know happened during the time of human evolution in East Africa.

"And continents drifting, changing their positions, mountain chains building up; the sea level rising and lowering. All these physical changes together, which actually express themselves locally in terms of climatic change, would be the initiator of evolutionary change."

Species can respond in three ways to these massive changes. They can move their habitat to a new geographical location consistent with the conditions of their former living area. Or they can become extinct, if for example there is no other habitat they can move to. Or finally, on rare occasions, genetic novelty will arise and a new species will be produced.

The involvement of environmental changes in the evolutionary process is not a new concept. Traditionally, environmental changes were part of a package of influences including competition and pure chance. Vrba's bold new idea is to "limit the initiation of the causal chain entirely to physical environmental changes. They are the *sine qua non* of the initiation of both speciation and extinction."

Peter Wheeler, an evolutionary biologist at Liverpool University who has made a special study of heat loss and retention, argues for a key positive advantage of standing upright. In hot conditions, especially when the sun is fully up, less of the body is exposed to direct sunlight, and accumulated heat in the body can be lost more quickly without consuming large quantities of water. Further, Wheeler thinks that two-legged locomotion, at least at low walking speeds, is less demanding of energy than its four-legged equivalent.

This ties in with Vrba's notion of the hominids having to cope with being in the open instead of the forest. The climate was cooling and drying enough to cause the forests to retreat, but the sun was very bright and the early hominids would have been directly exposed to it.

In hot, dry, open conditions, two-leggedness put the hominids at a distinct advantage, according to Wheeler: "Of all the animals on the savannah, we are the ones that are able to cope with the highest thermal stress . . . We keep the whole body cool, and we are able to protect our brains in that way."(Interview, September 1993.) And the ability to keep the brain cool, especially as it grew and became more powerful, was an indispensable key to human development.

These explanations, from Vrba and Wheeler, emphasize the climate and the environment. Another, partly contradictory view comes from the anatomist Owen Lovejoy. He thinks the riddle of upright walking cannot be solved unless a cause is found which fits the general behavior pattern of the animal, and he detects weaknesses in all the other arguments.

Sand dunes after a Sahara Desert rainstorm in Niger.

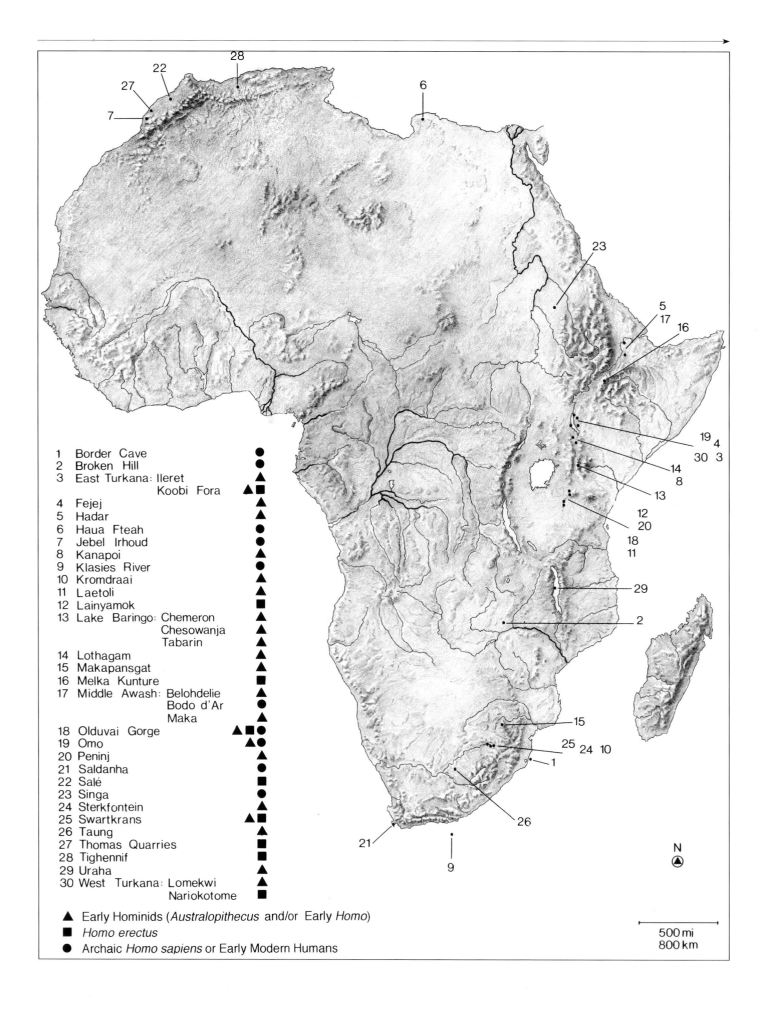

1   Border Cave ●
2   Broken Hill ●
3   East Turkana: Ileret ▲
        Koobi Fora ▲ ■
4   Fejej ▲
5   Hadar ▲
6   Haua Fteah ●
7   Jebel Irhoud ●
8   Kanapoi ●
9   Klasies River ●
10 Kromdraai ▲
11 Laetoli ▲
12 Lainyamok ■
13 Lake Baringo: Chemeron ▲
        Chesowanja ▲
        Tabarin ▲
14 Lothagam ▲
15 Makapansgat ▲
16 Melka Kunture ■
17 Middle Awash: Belohdelie ●
        Bodo d'Ar ●
        Maka ▲
18 Olduvai Gorge ▲ ■ ●
19 Omo ▲ ●
20 Peninj ▲
21 Saldanha ●
22 Salé ■
23 Singa ●
24 Sterkfontein ▲
25 Swartkrans ▲ ■
26 Taung ▲
27 Thomas Quarries ■
28 Tighennif ■
29 Uraha ▲
30 West Turkana: Lomekwi ▲
        Nariokotome ■

▲ Early Hominids (*Australopithecus* and/or Early *Homo*)
■ *Homo erectus*
● Archaic *Homo sapiens* or Early Modern Humans

N

500 mi
800 km

Chimpanzees can stand on two feet for long enough to reach food.

"Many times people will try to come up with a reason for walking upright which is singular. It's been suggested that it's a feeding adaptation, to pick fruit off trees, because chimpanzees stand bipedally in order to do that – but the fact that chimpanzees are fully capable of doing that and then once again return to quadrupedal walking, which is a very effective way to locomote, tells us that's not a good selective reason to adopt upright locomotion as a permanent adaptation."
(Interview, August 1993.)

He applies similar principles to other arguments. Seeing over the long grass is not valid, because upright walking probably began in the forest, not on the savannah. According to Lovejoy, there is a similar problem with the heat dispersal argument: upright walking began in the forest, when the heat of the sun could be more easily avoided.

The idea that bipedalism began before the retreat of the forests is based on Lovejoy's estimate that by the time Lucy was alive, the species was already very well established with that mode of locomotion, and that its starting-point must have preceded the climatic change which altered the landscape.

Lovejoy believes that the strongest pressure for change comes from reproductive behavior. Anything which increases the chance that babies will survive and thrive, will lead to a major new adaptation. And he believes that upright walking enabled males to gather high-protein foods from considerable distances and carry them back to their infants; the carrying was the key.

"Any behavior that would tip the balance and make those offspring more likely to survive would be strongly selectively favored. One thing that would do that would be if males were to become secondarily involved in the parenting process.

There is probably no single explanation for the
adaptation of standing and walking upright. All
the local circumstances must have played their
part, against the background of the need to gather
food and care for the young in an environment
where the landscape was becoming steadily more
open and less forested.

Chimpanzee infants cling to their mother's fur, using their hands and feet. Upright walking brings with it changes in the structure of the feet and hands, leading in turn to different ways of carrying babies.

"A way they can do that is occasionally to collect highly valued items of food, such as reptiles, amphibians, eggs, nestlings, grubs, worms. These items are extremely high in calories, in protein and in rare fatty acids and normally a hominoid must search for a long time to find them.

"If a male were to occasionally find one of these and provide it to the female it would greatly improve the time that she has available to care for the infant, it would raise the survivorship of the group and of the infant, and it would be an intensively favored form of adaptation. But that form of adaptation requires upright walking so that you can carry items like that."

(Interview, August 1993.)

And once you start walking upright you have to carry on, and get better at it. Infants, for example, lose the ability to grasp with their feet, so they have to be carried in the parents' arms rather than clinging to their fur. This in turn puts more pressure on the parents to walk consistently on two legs.

This begins to sound as though something resembling modern family units originated over three and a half million years ago, coinciding with the beginnings of bipedalism.

Social arrangements do play a major role in evolution. Much research has been done to explore how mating patterns and sexual behavior relate to reproductive advantage. In line with Lovejoy's ideas, many scientists have suggested that monogamy among humans came about because of the demands of child care, especially once larger-brained babies were being born which need so much care in their early years that the mothers need a devoted assistant.

On the other hand there is no biological reason, as Robin Dunbar points out in an article in the *Cambridge Encyclopedia of Human Evolution* (Cambridge University Press, 1992), why that assistant should be the child's father or indeed any other male. The assistant could just as well be the mother's sister or any other devoted friend.

And there is evidence to show that, in terms of evolutionary history, polygamy rather than monogamy has been the more usual human arrangement. The substantial difference in size between early male and female humans, and features related to sexual display in males, such as abundant body hair, are characteristics found in other species which are polygamous.

In the end any behavior which benefits reproduction and child survival will be likely to prevail. The point at which monogamy seemed to meet those needs is unclear, and it seems to stretch the imagination and the evidence too far to associate exclusive sexual partnerships with the gathering of food three and a half million years ago.

Lovejoy's scenario, describing the circumstances which may have brought upright walking into being, is imaginative; it paints a compelling picture of what life might have been like. On the other hand, it is based in part on his proposal of a very early date for the origin of bipedalism, an idea which is not widely shared.

One major strength of his argument is that no single explanation will do: the only way to begin to understand the concept of walking upright is to take account of the whole package: the climate, the changing environment, the social and family arrangements of the animals, the demands of child-rearing.

Robert Foley adds another element to the package, a reminder that evolutionary change does not happen on a global scale but locally, in response to very specific circumstances. Considering evolution from the perspective of millions of years, he says, can be very misleading.

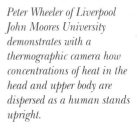

"Looking at the relationship between climate and evolution reminds us that it is the local, not the global environment that is important in the way evolution works; and so too it is necessary to remember that while evolution may occur over geological time, its underlying mechanisms – selection and drift – are played out every day and over very short time-spans. In reality the generation, not the geological epoch, is the key unit of evolutionary time." (The Curl Lecture, November 1993.)

Peter Wheeler of Liverpool John Moores University demonstrates with a thermographic camera how concentrations of heat in the head and upper body are dispersed as a human stands upright.

So the adaptation to upright walking may initially have taken place within a small community of animals in a particular clearing in the forest in response to social, feeding and reproductive pressures – against the background of climatic change rather than as a direct result of it.

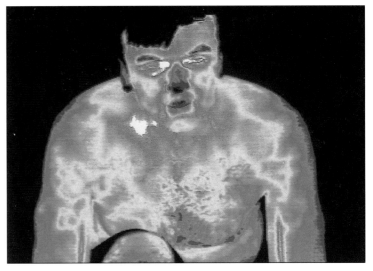

The evidence from Lucy and Laetoli is only twenty years old, and in that short time, scientists have had to try to make sense of the new information that walking upright was definitely the normal way for hominids to get around from as long ago as three and a half million years – and possibly a lot longer.

And yet whatever Lucy's skill on two legs, her brain was still ape-sized. The first hominids were upright apes.

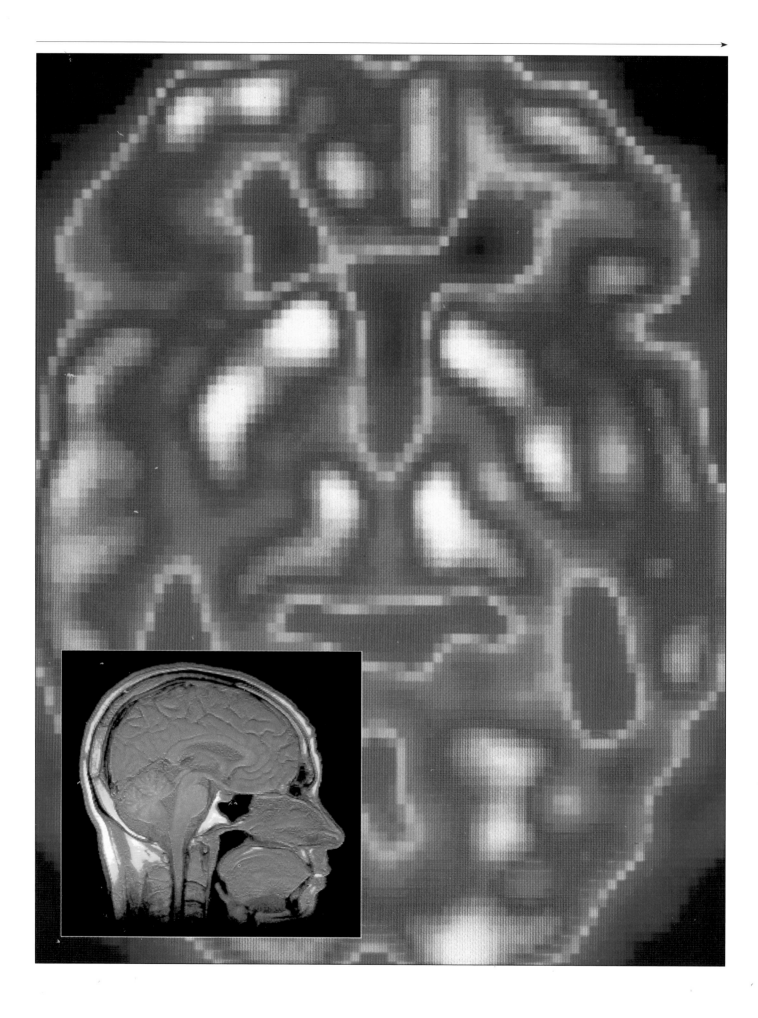

3 *Thinking*

The power of the human brain is another clear characteristic which appears to distinguish humans from other species. For the first hundred years or so of the science of human evolution there was no consensus about the order of events: did the development of the big brain drive all the other changes, such as upright walking, or did walking precede the growth of the brain?

The study of evolution is very young. Darwin published *The Origin of Species* in 1859. The first fossil discovery of an early human was made in Gibraltar in 1848, although it was not to be recognized as such for many years. In 1856 an ancient skeleton was found by workmen in Germany, and these bones, known as Neandertal man, were the first to open up the debate about the nature of prehistoric humans. Many of the most significant breakthroughs in knowledge and understanding have taken place only in the past twenty-five years, and a great deal of the previous century was taken up with the assertion of apparent facts and their subsequent revision in the light of new information.

The acceptance of the idea of evolution as distinct from religious creation is not as simple as it looks. Scientific inquiry is conducted by people, not by super-objective machines, and people carry with them the baggage of all their underlying assumptions and instinctive inclinations. Easy as it may be to say that it is overwhelmingly proven that humans are descended from apes, it is far from simple to exclude the continuing need to see humans as special, as intrinsically different from the apes.

It is very difficult for us to resist the temptation to put our existence on a different plane to that of other animals. Consciousness is a strange thing; no one has yet succeeded in defining it. One of its characteristics is that we are the only species fully aware of its own mortality. Other animals fear imminent death, and express that terror – we humans can daily contemplate a finite life, and it seems reasonable to assume that the knowledge of death (as distinct from the fear of death) gives us a very different attitude towards life.

Observing a scientist discuss the size and function of brains is fascinating. But it would take remarkable sang-froid not to feel disconcerted at the sight of the scientist demonstrating his information by handling, without reverence and in no special order other than size,

Gibraltar man was found during the construction of military fortifications. It is not known who found the skull, or even exactly when or where.

OPPOSITE: *A scan created with a radioactive tracer, injected into the bloodstream, to reveal metabolic activity in the brain. Blue shows low activity, and yellow shows high activity. The normal pattern is roughly symmetrical between the left and right hemispheres. Inset, a false-color nuclear magnetic resonance image of a front-to-back section through a human head, showing the structures of a normal brain.*

the brains of whales, dolphins, humans, chimpanzees and gibbons. As it happens, the human brain is in the middle of the range; it does not belong in a separate category, to be laid on a different table because it is the very centre of everything which enables us to be human.

If our ancestors came down from the trees before they developed their big and powerful brains, then it seems we are somehow closer to the apes than if it were the case that the big brain came first and long ago, driving all our subsequent development. To be an upright ape before becoming an upright human emphasizes "ape-ness" over "human-ness."

In his book *Bones of Contention* (Penguin, 1991), Roger Lewin quotes Henry Fairfield Osborn, Director of the American Museum of Natural History, as telling a scientific audience in December 1929: "To my mind the human brain is the most marvellous and mysterious object in the whole universe and no geologic period seems too long to allow for its natural evolution."

And a couple of years before, he had written in a scientific journal: "The most welcome gift from anthropology to humanity will be the banishment of the myth and bogie of ape-man ancestry and the substitution of a long line of ancestors of our own at the dividing point which separates the terrestrial from the arboreal line of primates."

He seems to be saying that it is acceptable to be descended from apes, to be a primate, as long as the separation took place a really long time ago, and as long as the human brain can have some unique characteristic which makes it different from the organ which occupies the skulls of other, ordinary animals.

It is easy to poke fun at extreme examples like Osborn's comments – until one remembers that it was not until 1953 that the notorious Piltdown skull was exposed as a hoax. Piltdown's main feature was that it combined a human-like brain case with an ape-like jaw, suggesting that the big brain preceded and therefore drove other, later changes. Piltdown was largely accepted as a genuine, if slightly puzzling, fossil for forty years. The scientists observing it and writing about it were not stupid or half-blind. There has to be some other reason for the collective failure of the community to denounce it as a rather incompetent fake – and the reason has to lie somewhere in the notion that a very early big brain was a good and reassuring thing to believe in.

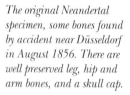

The original Neandertal specimen, some bones found by accident near Düsseldorf in August 1856. There are well preserved leg, hip and arm bones, and a skull cap.

The *Illustrated London News* trumpeted the discovery of Piltdown in its edition of 28 December 1912: "A discovery of supreme importance to all who are interested in the history of the human race was announced at the Geological Society when Mr Charles Dawson of Lewes and Dr A. Smith Woodward, the Keeper of the Geological Department of the British Museum, displayed to an eager audience a part of the jaw and a portion of the skull of the most ancient inhabitant of England, if not in Europe."

The remains had been found the year before in a gravel pit close to Piltdown Common in Sussex, England. The account in the *Illustrated London News* is a fascinating evocation of thinking at the time. W. P. Pycraft wrote that the evidence was "incontrovertible" that Piltdown represents "a link with our remote ancestors, the apes."

The lower jaw, he tells us, is unmistakably ape-like in its structure; the teeth themselves were more human – "there is reason to suspect that the canine or eye teeth projected . . . slightly above the level of the rest – an ape-like character met with in savage races today, though never to the same extent as in apes."

The upper part of the skull, on the other hand, is human. Pycraft writes that it had a brain capacity of "just under two pints, which is nearly twice as much as that of the highest apes." The brain size delivers the fact that the find was a human ancestor; the jaw and teeth mean he was a link with the apes of the past.

Pycraft goes on to tell us that the Piltdown Man probably lived "several hundred thousand years ago, perhaps a million," and then paints a charming picture of his life. He was a man of low stature, not yet having achieved the "graceful poise of the body which is so characteristic of the human race today." But he was by no means lacking in intelligence. "Living in a genial climate amid a luxurious vegetation, and surrounded by an abundance of game, he may be said to have led a life of comparative ease. Of clothing he had no need; nor was there any need to bother much about housing accommodation."

On December 28 1912 the Illustrated London News *reconstructed an image of what the complete Piltdown Man may have looked like: "the most ancient known inhabitant of England." Inset is a model reconstruction of the skull itself: the "fossil" parts are dark coloured, with the rest of the skull filled in.*

The Piltdown site, where a goose owned by the local squire is said to have been present every time digging took place – at least until the death of Charles Dawson (second left), after which it stopped coming along. Arthur Smith Woodward is on the right.

Apparently he hunted and ate elephants and rhinoceroses (what with, Pycraft does not tell us), while dodging the attentions of lions, bears and sabre-toothed tigers. For this last purpose, "he may have been forced to devise some kind of shelter by night."

Running through this account is a sophisticated individual, someone whose ape-like features do not get in the way of his human characteristics. This of course allows a sort of sanitized evolution, in which the elements of the anatomy inherited from the apes are neatly tucked away in the jaw while the brain gets on with the serious business of being a "proper" human being, already bright enough to kill and dismember elephants with bits of broken flint and with time on his hands to enjoy the pleasant climate of the south of England.

Piltdown Man took his place as a human ancestor alongside fossils previously discovered in Europe and Asia. Nothing at that point had been discovered in Africa; the home of humanity was generally believed to be in Asia. The English scientific establishment was very proud that Piltdown could join the existing tiny collection of fossils from places as far apart as Java and Germany; national pride was satisfied by the discovery of such an important English specimen, especially one which carried a highly significant message.

Piltdown had been discovered by Charles Dawson, a lawyer and amateur archaeologist. He took it to Arthur Smith Woodward, Keeper of Geology at the British Museum (Natural History), who helped him with further excavations and finally gave the remains his seal of scientific approval. Further powerful support came from Arthur Keith, Conservator of the Hunterian Museum at the Royal College of Surgeons. And more endorsement still was provided by Grafton Elliot Smith, Professor of Anatomy at Manchester University. These three scientists, all later knighted, were effectively unchallengeable. If they said Piltdown Man was genuine, and that he demonstrated the primacy of the brain in human evolution, then the case was for all practical purposes closed.

Closed, but with dissenters. In the *American Museum Journal* in 1914 William King Gregory published an article concluding that Piltdown Man may have been genuine but reporting the distinct possibility that he was not. "It has been suspected by some," he wrote, "that geologically [the specimens] are not old at all; that they may even represent a deliberate hoax, a negro or Australian skull and a broken ape-jaw, artificially fossilized and 'planted' in the gravel bed, to fool the scientists." Sadly, he was too discreet to name the "some" who suspected fraud. If the suspicion was widespread, it was nevertheless effectively silenced, presumably by a process of self-censorship.

Piltdown had just too much going for it: the support of the key figures of the British scientific world, the happy fact of it being a specifically British contribution, and, of course, its confirmation of the "brain-first" theory of evolution. In any case, other world events in 1914 would have made it difficult to accuse important Englishmen of deceit and trickery.

The rumours reported by William King Gregory were completely accurate, however. Piltdown consisted of a recent human skull (probably no more than 500 years old), together with the jaw-bone of an orang-utan of around the same vintage. They had been treated to look like fossils, tell-tale parts had been removed and the teeth filed down; they had been placed at the appropriate level in the gravel pit alongside imported mammal fossils from the correct period in order to substantiate the dating. This was no case of mistaken identity or wishful thinking. It was an absolutely deliberate and careful fraud.

It was not exposed as such until 1953. Tests established the real dates and origins of the bones, and Piltdown was deposed. The perpetrators were never found, however; no one confessed, and no "smoking gun" evidence has turned up. Charles Dawson, the original "finder," has long been suspected; Phillip Tobias in South Africa believes that he received authoritative scientific help from among the senior British figures who endorsed the find.

One of the difficulties about Piltdown is working out the level at which it was a deceit. It would be one thing for a major scientist to manufacture a fake skull with the deliberate

A painting by John Cook of all the actors in the Piltdown affair. Arthur Keith is second from the left in the front row, with W.P. Pycraft on his left. In the back row, from the right, are Arthur Smith Woodward, Charles Dawson, and Grafton Elliot Smith.

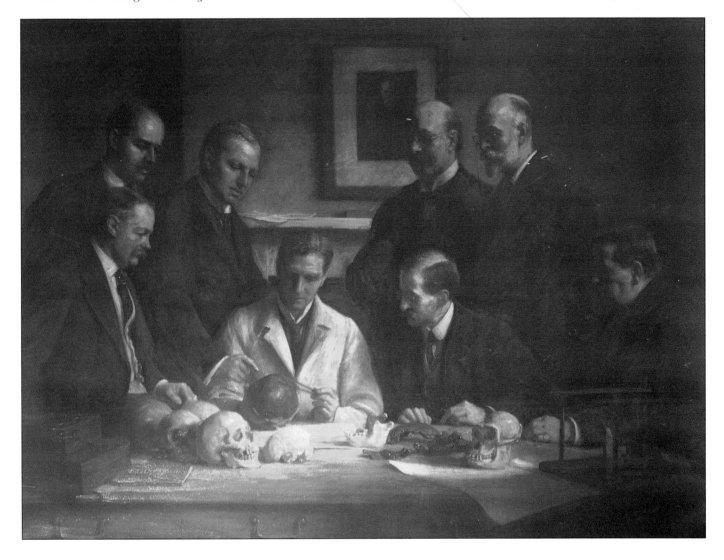

The Taung skull, found by Raymond Dart in South Africa in 1924, was ridiculed because it flatly contradicted Piltdown. It had reduced canine teeth, a flattish face, and signs of upright posture – but, unlike Piltdown, a small brain.

intent to provide "evidence" in support of his own theories. It would be another if the faking were done by a student, perhaps as a joke; and one can imagine the problems that would later arise once the joke had been accepted as genuine by the most senior scientists in the land, books had been written about it and whole academic careers founded upon it. It would be problematical, to put it mildly, to admit to the prank once matters had proceeded that far.

The point about Piltdown, whatever its source, was that it fitted in with the ethos of the time. Scientists were not surprised to find apparent evidence of the early development of the big brain, because, in the state of knowledge of the time, that is what they thought had happened.

If it took forty years to expose the hoax, it took far less time for other fossils to come to light which appeared to contradict Piltdown's message. But such was Piltdown's primacy, and the authority of its advocates, that the new finds were challenged or ignored. In fact, Piltdown set back the cause of the understanding of human evolution by nearly half a century.

In 1924 Raymond Dart discovered a tiny fossilized skull at a place called Taung, near Kimberley in South Africa. Dart was a young Australian, thirty-two years old at the time of the discovery, and he had a reputation for being ebullient and heretical. He represented the very antithesis of the stuffy English establishment which had launched the Piltdown skull on the world only ten years before.

Ironically enough, Dart had been a student of Grafton Elliot Smith in England, and had been sponsored by his teacher in his move to the Department of Anatomy at Witwatersrand University. At the time, South Africa was a scientific backwater; nothing bearing on human evolution had been found there, and nothing seemed likely to be found – research was centred on Europe and Asia.

Dart thought that the Taung skull, which was about two million years old, showed an unprecedented blend of ape-like and human-like features. In particular it had very small canine teeth; no bigger than the canines of modern human children, whereas in the young of gorillas or chimpanzees, the canines are large protruding fangs. Dart also noticed that the face was very nearly vertical, whereas the faces of ape children have a snout.

The only other "intermediate" fossil available at the time, showing features of both ape and human, was Piltdown. The Taung baby, as it came to be known, was the precise reverse of Piltdown: a brain the size of an ape's, combined with many human-like structural features, most particularly the teeth and the shape of the face. It also seemed to Dart that the point at which the spine would have joined the skull suggested upright walking.

The brain case itself was also remarkable. The process of fossilization had created an immaculate impression of the brain itself inside the skull, where lime-rich sand had accumulated, solidified and taken a print, as it were, of the indentations left behind by the brain.

Phillip Tobias, who today holds the Chair of Anatomy at Witwatersrand University formerly held by Dart, and who describes Dart as "my predecessor, my mentor, my very dear friend over many years," takes up the story.

"One doesn't often realize how soft bone is, that it can be moulded and take impressions, so that when one looks at this, one can see the impression even of some of the little convolutions of the brain. One can even make out the impression of an artery running over the cast.

"Dart was a student of the brain and he thought he could detect certain features in the back part of the brain cast, which were human-like rather than ape-like, so he was confronted with a creature whose total brain size was no bigger than an ape yet whose brain form was more like that of a human being.

"He took the great jump of concluding that this child was a representative of a creature that was knocking on the door of humanity, but hadn't quite crossed the threshold of humanity." (Interview, August 1993.)

Dart published his findings very quickly and he gave his find a new name – *Australopithecus africanus*. *Australopithecus* means "southern ape," thus distinguishing the creature from our own genus *Homo*, while the species name *africanus* reflects its African origin. Phillip Tobias comments on the response from the establishment.

"It was a shocking discovery. The world wasn't ready for it, and for twenty-five years the world did not accept what Dart claimed for it. But he looked up Charles Darwin's 1871 book, *The Descent of Man*, and there Darwin stated that if ever we are to find an ancestor of human beings, it is rather more likely that it will be found in Africa than anywhere else in the world, and Dart said, I have confirmed Darwin's prediction." (Interview, August 1993.)

The Illustrated London News *published a front-page reconstruction of the Taung baby (above) on February 14 1925, describing its picture as showing "an incipient sense of humour, and a dawning light of intelligence."*

Phillip Tobias (below) has the original fossil in his safe at Witwatersrand University in South Africa.

Early hominids such as
Australopithecus africanus
were the frequent victims of
predators like leopards and
sabertooth tigers. Only very
much later did hominid
species start to dominate other
creatures.

Phillip Tobias is pictured (left) with his predecessor as Professor of Anatomy at Witwatersrand University, Raymond Dart. Robert Broom (inset) found an australopithecine fossil skull (below) at Sterkfontein in 1947, which was known as "Mrs Ples," and which confirmed Dart's view of the place of the Taung baby in human evolution.

It was a very long time before Dart's discovery was accepted, partly because it simply failed to fit what the established scientists expected to find. They thought the cradle of humanity was Asia; they thought the big brain came before, or at worst was simultaneous with, bipedalism – and they were frankly prejudiced against the very idea of human origins being in Africa. To the Europeans of the day, the idea was racially and culturally repugnant.

Dart belonged most emphatically to the group of scientists willing to announce their conclusions and await development. Tobias describes him as being one of those who "erect signposts twenty miles down the road," as against those who "come along in their wake, filling out the details, a centimetre at a time, inching their way towards the signposts." It was a quarter of a century before the rest of the world reached Dart's signpost, and not before he had, for several years, given up the whole study in disgust at the isolation he felt.

The Taung baby was in fact a child of between four and six years old. One of the more rational arguments used to reject Dart's finding was to do with the fact that the Taung specimen was so young.

Terrence Deacon, a neurobiologist at Boston University, points out that all ape-children appear somewhat human-like at a very early age – only as they grow older do apes take on their characteristic appearance. But other, later finds in South Africa supported the Taung baby's place in the record, most particularly an adult fossil skull found at Sterkfontein by Robert Broom in 1947, and nicknamed "Mrs Ples."

She was placed in the same species, and hardly anyone now challenges the identification of the Taung child as an australopithecine, dating at about two million years ago. Dart is finally recognized as having permanently changed the study of human evolution. The Taung skull proved, fifty years before the discovery of Lucy, that upright walking did indeed precede the development of the large brain.

Humans do not have the largest brains of all species in absolute terms. Whales and elephants have bigger brains; dolphins' are similar in size to humans'. The average modern human brain has a capacity of about 1,350 cubic cm, whereas a big whale's brain can be four times that size. Whales are not four times as bright as humans, however, so absolute size is obviously not the governing factor in providing a guide to the brain's power.

What does seem to be significant is the relative size of the brain when compared to body weight. Primates have larger brains for their body size than most non-primate mammals, and modern humans have the largest brain, in relative and absolute terms, of any primate that has ever existed. (Terrence Deacon, *The Cambridge Encyclopedia of Human Evolution*, Cambridge University Press, 1992.)

Humans not only have big brains relative to their body weight; their brains grow to a great extent as they mature. Humans actually have a gestation period of twenty-one months, not nine; babies' brains continue to grow at the pre-birth rate for a further twelve months after they are born. (Roger Lewin, *Human Evolution, An Illustrated Introduction*, third edition, Blackwell, 1993.)

The fossil record tells us a good deal about brain size. Lucy's brain (*Australopithecus afarensis*) measures about 400 cubic cm, which is about the same size as that of a modern chimpanzee. She had approximately the same body size as a chimpanzee as well.

The brain sizes of the various species identified as *Homo* show clear growth. *Homo habilis*, dated to about two million years ago, had a brain capacity of 650–800 cubic cm; *H. erectus*, dated to about one million years ago, had a brain of 850–1,000 cubic cm; early *H. sapiens* and Neandertals had brains ranging from 1,100 to 1,400 cubic cm.

But the relatively small size of Lucy's brain – 400 cubic cm – coexisted with fully upright walking, providing further proof, consistent with the Taung skull, that the large brain came after, and not before, bipedalism.

What does the big brain actually do? What is the point of having it? What are the mechanics of how it works? How does it contribute to *Homo sapiens*' specialness? And when did it develop?

This is a very complicated subject indeed. The physical structure of the brain is not fully understood; it is not like a consciously designed computer in which you can identify and label all the wires and switches, and say with certainty what each component does. It is not even possible to say that a particular section of one species' brain has the same function as the equivalent in another species' brain.

ABOVE: *A model reconstruction of an adult female* Australopithecus africanus *such as "Mrs Ples".*

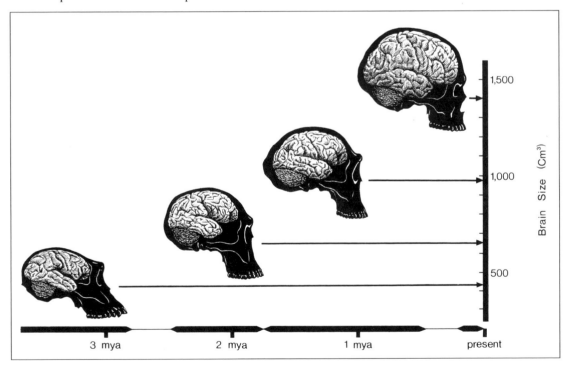

LEFT: *The chart shows the changing brain size of, from the left,* A. afarensis, Homo habilis, H. erectus, *and modern humans (*H. sapiens*).*

And even if this were the case, there remains a function of the brain which no computer has yet been able to reproduce: the ability to respond laterally, unpredictably and creatively. You still have to tell a computer what to do; brains do things on their own. The independent creativity of the brain enables humans to do things spontaneously. It is at the root of our ability to think, to plan and drastically to affect our environment (for good or ill), and therefore, as it were, to begin to control our own and other species' evolution. The biggest and most sophisticated computer in the world still needs software made by humans before it can do anything. The human brain comes ready-equipped with both hardware and software.

There are no fossil brains. That part of the body is never preserved by time. The size and shape of the skull provide pointers as to what was probably inside it; but the actual brains have disappeared. Sometimes, as happened with the Taung baby, the surface of the brain makes indentations on the inside of the skull, and that in turn leaves outlines on the material which fills up the empty skull after death; but these endocasts, as they are called, can only give information about the surface of the brain – nothing can be deduced about its contents.

Specialists in this field can therefore only examine the skulls of fossil remains, and the brains of living creatures. Terrence Deacon of Boston University and Harvard Medical School says: "A lot of my interest in the field has been the study of the development of brains from embryos to adults, to find out what the principles are by which brains get built. From that you can infer, with comparative evidence from other species, and a little bit of the evidence from fossils, exactly how evolution could have modified brains over the course of time – because those are the tools that evolution would, in a sense, have access to."

Deacon describes himself, very vividly, as a "wet biologist." He studies brains by picking them up, handling them, taking them to bits to see how they work. At the core of his "wet biology" is the fact that human brains have no new component which evolution has "added" which would account for human special abilities. The fact that a human brain is bigger than a chimpanzee's does not mean that it has an extra part which would explain why humans can do things chimps cannot. Nor, on the other hand, is a human brain simply an expanded version of a chimp's brain, in which all the bits are equally bigger, work better and cumulatively improve the overall power of the organ.

Petrified sediments can fill the skull after death, recording an exact impression of the inside of the skull – the nearest thing to fossil brains. These australopithecine brain casts (above) were found at Sterkfontein in South Africa in 1936. To the right are reconstructions of a range of hominid skulls: in chronological order from left to right, Australopithecus, Homo habilis, H. erectus, *Broken Hill Man (a specimen usually described as archaic* H. sapiens), *Neandertal man, and modern* H. sapiens.

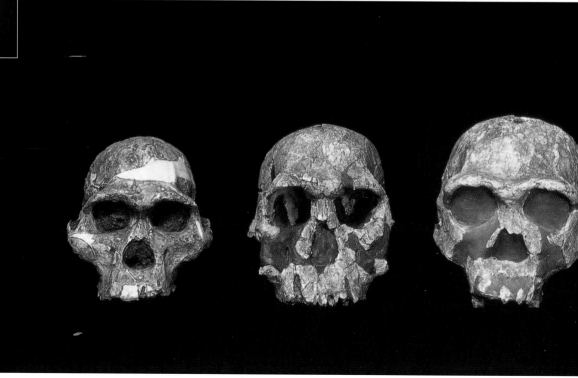

"It turns out," says Deacon, "that as we've studied the comparative anatomy of brains, for the most part they are very conservative. They have a lot of common structure.

"There really isn't a part of the brain that is human only. Most of the differences have to do with proportions, relative sizes of parts, and the effects that those differences in size have on connection patterns and finally on function."

The covering on the surface of the human brain is large in relation to the rest of the brain, as is a section of the covering right at the front known as the pre-frontal cortex. Deacon says it is about twice as large as it would be in a typical primate brain of around the same overall size. And in absolute terms, it is about six times the size of the same part of a chimpanzee's brain.

The changes in the relative sizes of parts of the brain take place as humans grow from birth. Even in human foetuses the brain covering, the cortex, is relatively large; but once a child is born, the pre-frontal cortex develops still further. The growth of the human brain cannot take place inside the womb – this need for further growth creates a lengthy period of helplessness in human babies.

In all mammals, the pre-frontal cortex handles sequences of events. It works out relationships between activities. Any mammal with damage to that part of the brain becomes very confused and impulsive; it loses the ability to analyze sequences, where one thing depends on another. It cannot stop doing things, even though it is obvious that it should want to.

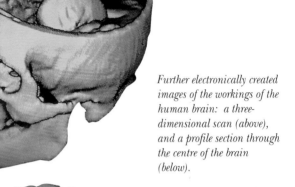

Further electronically created images of the workings of the human brain: a three-dimensional scan (above), and a profile section through the centre of the brain (below).

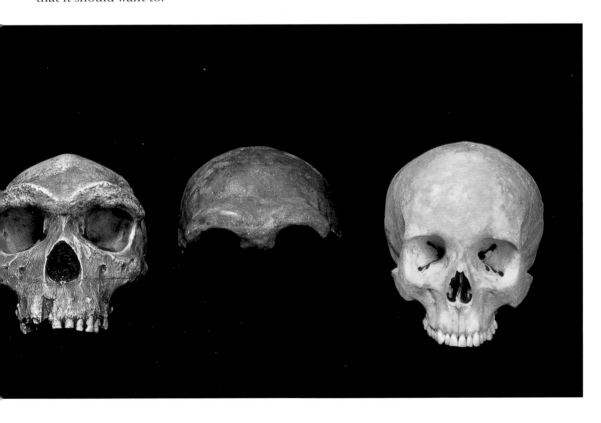

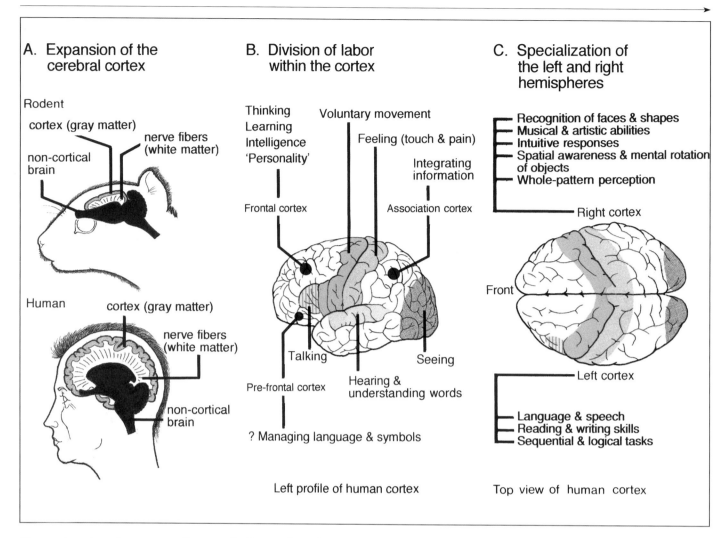

A. Expansion of the cerebral cortex

Rodent

cortex (gray matter)

nerve fibers (white matter)

non-cortical brain

Human

cortex (gray matter)

nerve fibers (white matter)

non-cortical brain

B. Division of labor within the cortex

Thinking
Learning
Intelligence
'Personality'

Voluntary movement

Feeling (touch & pain)

Integrating information

Frontal cortex

Association cortex

Talking

Seeing

Pre-frontal cortex

Hearing & understanding words

? Managing language & symbols

Left profile of human cortex

C. Specialization of the left and right hemispheres

Recognition of faces & shapes
Musical & artistic abilities
Intuitive responses
Spatial awareness & mental rotation of objects
Whole-pattern perception

Right cortex

Front

Left cortex

Language & speech
Reading & writing skills
Sequential & logical tasks

Top view of human cortex

Diagram A compares the overall structure of the human brain with that of a typical mammal. Note the increased proportion of cerebral cortex in the human. Diagram B indicates how the cortex is divided up for various functions. Diagram C shows the different specialized functions of the left and right hemispheres of the brain. This specialization does not occur in chimpanzees; in humans,the asymmetry between the two sides develops differently in males and females.

Deacon believes that the pre-frontal cortex in humans is able to drive the rest of the brain, or has recruited it to a very specialized function – the processing of symbols and language. In humans, many parts of the brain, which in other primates are used for other purposes, are drawn into the manufacture of language and communication.

It does not seem to be possible to say that a particular part of the human brain is the bit that "does" language. Instead, many parts of the brain, aided and abetted by the crucial pre-frontal cortex, are drawn into creating this one vital activity. Deacon's studies suggest that language ability is dispersed over different parts of the brain, principally but not exclusively on the left hemisphere. The parts of the brain involved in language even seem to vary from language to language. Individuals who speak several languages appear to draw on different parts of their brains for each one. And the parts of the brain involved in language include areas which in other primates do other things.

"What's happened is that they've become recruited for a new function," Deacon says. "In some ways it's a sort of software revolution. The same kind of computer is being used for a very different function."

Deacon thinks that the actual size of the brain is not exactly the point. The human brain is indeed relatively large; but its most interesting feature is the relative size of its component parts, and the important issue is what brought about the changes in those relative sizes. Deacon has a clear answer to that puzzle.

"My own view is that the human brain is a consequence of the demands that language placed upon it. I would place the origins of language back at the very beginning of when we see changes in human brain size. Brain size is a good indicator that the restructuring has begun inside; as soon as we see that in the fossil record, I think we can infer that the demands of language have already been having an effect on brain structure.

"Language in some form probably began somewhere in the range of two million years ago, with a creature like *Homo habilis*, and continued to play a role in altering human social behavior, human activities and the human brain for a million and a half years or so from that time." (Interview, August 1993.)

By this Deacon does not mean that hominids two million years ago were using language in the way we do today; simply that by that time it was possible to refer to things in the abstract, to use words, signs or gestures to convey ideas other than straightforward practical information.

This opinion, that language in its simplest form was the sole or major driving force for brain development, is not universally held. Philip Lieberman, of Brown University, Rhode Island, takes a slightly different view. He thinks that complex language depends on having a more complex brain, and he points out that other anatomical adaptations necessary for speech, like the vocal tract, would be useless unless the brain already had the computing power to drive language. Language is software, useless without the hardware to drive it. And different software could be used to deliver abilities other than speech by the developed brain.

There is a relationship, Lieberman believes, between the parts of the brain involved in speech and the parts concerned with detailed, fine control of hand actions. The manufacture and use of stone tools, the first form of human technology, dates back to over two million years ago. Making even the simplest tools, essentially by bashing one piece of stone against another to achieve the desired shape, obviously requires careful control of the hands and arms and a degree of knowledge and planning.

The excavation site at East Turkana in northern Kenya. Richard Leakey started working there in 1968, and in 1972 he found the famous Homo habilis *fossil, KNM-ER 1470.*

Ape Man

At an excavation site at Olorgasailie, not far from Nairobi in Kenya, clear signs have been found of the butchery of elephants by early hominids. Microscopic examination of cut marks on the bones show them to have been made by stone tools. It can only be assumed that the elephant had died of natural causes, or as a result of some accident, and that the hominids scavenged the carcase for everything that they could cut away.

Homo habilis was probably the first hominid species to use stone tools to get at food. The tools, sharpened flakes or axes, made possible the swift and efficient butchery of animals; the meat was then taken to a safe place where it could be eaten out of the way of other predators in search of the same food.

The early hominids were far from being the dominant species that humans are today; they would have been constantly at the mercy of other wild animals on the open savannah. By any standards, they were a vulnerable and relatively fragile species. There would have been enormous pressure on them to develop new ways to survive. The use of tools would have made a very big difference, principally in terms of speed.

Many scientific debates have taken place about the diet of the early hominids. It seems likely that the earlier species, the australopithecines, ate vegetable foods (though not exclusively), whereas by the time *Homo habilis* and the slightly later *H. erectus* come along, a decisive shift had been made towards meat.

Not that there is any sign, as far back as one or two million years, that the *Homo habilis* and *H. erectus* were hunting in the sense that we understand it. They may have managed to catch and kill small, weakened animals, but for the most part it seems that they were scavenging on the carcasses of animals which had either been killed by other predators or had died by accident.

There is a clear relationship between the making of the tools and the purpose for which they were used. Meat and other animal foods like bone marrow are highly nutritious, and the growth of the brain demands substantial quantities of high-quality, readily digestible foods. In the absence of organized hunting, stone tools, as we have seen, would have enabled the hominids to scavenge their meat fast and effectively.

Meat had two advantages. Not only was it highly nutritious, it was also readily available across very broad geographical areas. Vegetable-based diets tend to restrict those who depend on them to the specific area where a particular food grows. Meat diets allow far more areas to be exploited.

Cheetahs are naturally able to chase down and kill their prey, a skill well out of reach of the early hominids, who had no hunting weapons with which to tackle fast-moving animals.

Stone tools of the kind found in the Olduvai Gorge in Tanzania, dating to slightly less than 1 million years ago. Above, a handaxe, and on the left, handaxes and cleavers.

Quality of food, as previously mentioned, is one of the keys to brain development. In one sense the brain is just another organ; biologically it is a perfectly ordinary part of the body which needs fuel like any other. It so happens, though, that as the brain expands it demands disproportionate amounts of good food. Phyllis Lee, a primatologist at Cambridge University, describes the chain of events: "What permitted us to have a big brain was the ability to have very high-quality foods. If you want to have a big brain as an adult, you have to do this by extending the period of brain growth in childhood."
(Interview, September 1993.)

There comes a point in the development of hominid or human babies where their brains can no longer grow in the womb. If they did so, the baby would quite simply be too large to be born. So the brain continues to grow and develop after the baby has left the womb, and the growth has to be rapid because at that stage the baby is acutely vulnerable. Phyllis Lee continues: "You have to be able to fuel the increase in brain size, you have to be able to feed a baby the kind of food that will allow its brain to grow very, very rapidly.

"Why increase the chances of a baby dying when really you should be decreasing those chances? There must be some very good reason why we need a very large brain. And there are many different reasons that people have put forward as to why we need very large brains, ranging from the ability to harvest foods to the need to cope with an incredibly complex social world. And we have all the technological world that goes with *Homo erectus*, of fire, of complex tools, all of which create environmental stability within a context of great social change."

These changes demand that individuals make relationships with one another, working out who is a friend and who is an enemy and how to live effectively in large groups. "This kind of knowledge requires a very large brain and very complex kinds of relationships, and probably complex forms of communication as well."

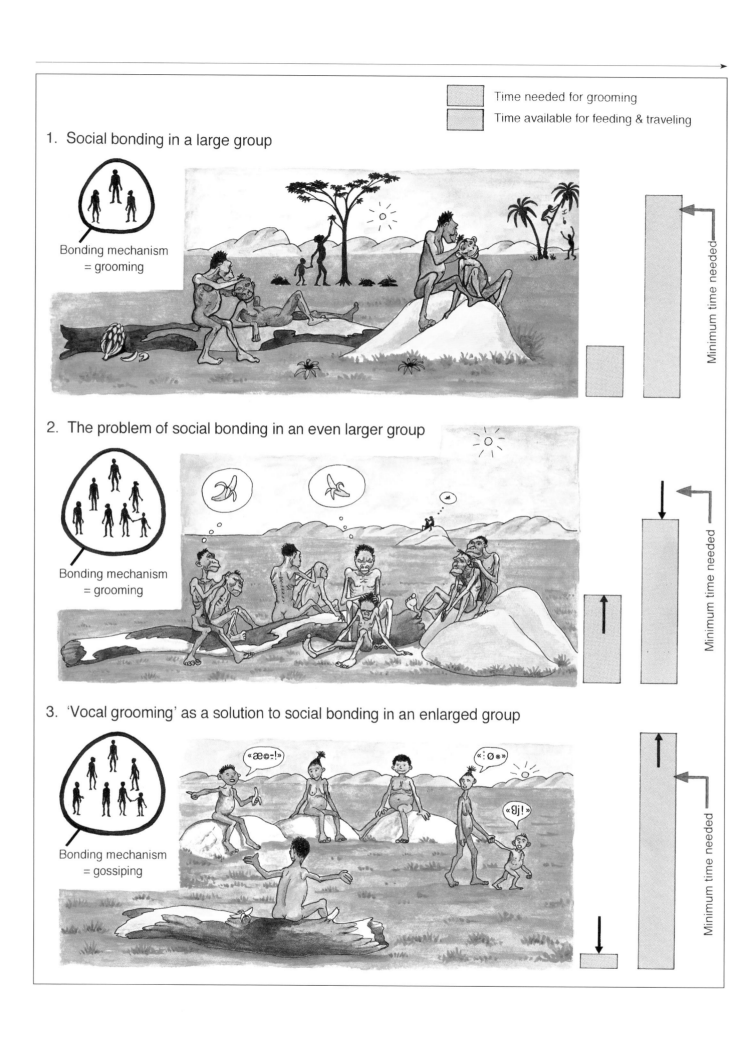

The notion of the connection between the big brain and the groups in which creatures live is reinforced by Robin Dunbar, a primatologist at University College, London. He believes that the larger the groups in which primates live, the larger their brains.

Small-brained monkeys live in small groups; baboons live in groups of forty to eighty individuals; chimpanzees, with larger brains still, relative to their body size, in groups of fifty to 100. The larger the group, the more difficult it is for each individual to work out how to deal with all the others. "The more individuals you have, the more possible combinations of who's friends with whom. You need a bigger brain to handle the increasing complexity of the relationships within the groups."
(Interview, September 1993.)

The quality of the relationships changes as well. "You get increasing use of more and more sophisticated kinds of social strategies. Chimpanzees can exploit social knowledge to deceive each other, to get each other to do things, to play politics, in ways which baboons are quite good at but nowhere near as sophisticated."

Primates conduct their personal relationships through grooming one another. Plainly, the bigger the group, the more time has to be spent every day grooming – and there will come a point where the needs of grooming and the needs of finding and eating food will clash. There is a time constraint on the possible size of the group; beyond a certain number of individuals, there are simply not enough hours in the day to attend to the social needs of grooming and the practical needs of feeding.

Ecological pressure was driving the early hominids into living in larger and larger groups, and this change, according to Robin Dunbar, was one cause of the increase in brain size, leading in turn to new ways of building relationships which transcended the restrictions placed on group size by the needs of grooming on a one-to-one basis.

Robert Foley stresses the interdependence of all the factors involved in the development of the brain. "The causes of large brains in terms of selective pressures may . . . be said to be social. Social complexity should drive brain size up. However, whether animals live in social groups or not is a function of various factors, such as body size, predation pressure, food distribution, and so on.

"Beyond this, brains are metabolically very expensive tissue, and therefore no matter how beneficial the presence of a large brain may be, the energetic costs may inhibit its selection."
(Curl Lecture, November 1993.)

Once brains begin to grow in size, they slow down growth rate and increase demand for high-quality foods. The growth of the brain is indeed an expensive and demanding process, which may explain why it has only occurred in hominids; only they were in a position to develop the circumstances to make it possible.

Around two million years ago, the first big-brained, tool-using hominids began to thrive in Africa. The development of the brain, a major key in the evolutionary story, seems to have been affected by several related changes. The way the hominids were living, the mental and physical equipment they began to use to compete with other creatures, the food they ate and the way they communicated – all these factors converged in the brain.

OPPOSITE: *The larger the group, the harder it is for its members to bond with one another through grooming alone. At a certain group size, time simply runs out. Language makes larger groups of individuals possible.*

BELOW: *A group of baboons play-fighting in the Kruger National Park in South Africa.*

4 Talking

Virtually all creatures communicate with one another. Birds sing; chimpanzees chatter; whales sing; insects seem to have some way of letting one another know what to do next and where to go. But humans have language, an intrinsically different concept. Language involves the infinite arrangement and rearrangement of precise and subtle sounds into different orders conveying different meanings.

The rules of language – grammar and syntax – mean that it is a variable instrument. The sounds we make can have changed meanings according to the context in which they appear.

Language is not the same as other forms of communication. Understanding when and why it began provides a pointer to one of the most significant periods in human evolution. Language is a tool for communicating, but it is also inextricably bound up with thought, the expression of ideas and creativity.

Robin Dunbar at University College, London says that language, the sophisticated speech we use every day, is unique to the human species. "There are still some doubts about dolphins and porpoises, but for all other primates and all other land-based animals, although they show some of the roots of language in very, very primitive form, they are nowhere within the same league as living humans in terms of the complexity of information which they can exchange by language."
(Interview, September 1993.)

Other animals can be taught to recognize and even imitate human speech. Dogs can "learn" about twenty or thirty words, and parrots can reproduce them. But no dog can voluntarily talk; and parrots do not appear to distinguish between copying words and imitating other sounds like circular saws. The process of voluntary speech is simply different.

Chimpanzees and baboons grunt and call to one another, and the sounds they make convey information. They can recognize each other's "voices," they can detect each other's moods and act on information they are being given – warnings about predators, news about food, or information about a move to another part of the habitat. "I'm going over there: give me a shout if you see any leopards." The sounds are simple, but they can be more than mere instinctive grunts of fear or happiness. They seem more human than the sounds made by non-primates.

ABOVE: *Like many animals, baboons communicate through threatening displays.*

OPPOSITE: *If the hominid lineage stretched back to* Ramapithecus, *a 15 million year old fossil found in Pakistan in the 1930s, the origin of language could be very early. But* Ramapithecus *was an ancestor to orangutans and the hominid lineage is no more than 8 million years old.*

There have been many studies of communication in the non-human primate world. Jared Diamond, in his book *The Rise and Fall of the Third Chimpanzee* (Vintage, 1991), quotes extensive research into the communicative abilities of the vervet monkey. The research work was done by Robert Seyfarth and Dorothy Cheney, beginning in 1977, in the Amboseli National Park in Kenya, and they published an account both of the vervets and of other animal communication in 1990 (*How Monkeys See the World*, University of Chicago Press, 1990).

The vervet has a vocabulary of about ten different sounds, some of which seem to be used deliberately rather than just as an involuntary response to some external stimulus. The sounds are more than just an uncontrolled expression of fear such as any animal – even a human – might make when threatened by something really scary.

Vervets use the grunt meaning "leopard" when there is a leopard, or other similar-looking and dangerous cat, in the vicinity. But they apparently only use it when there are other vervets around to hear the message and take appropriate avoiding action. A vervet on its own can be chased about by a big cat for some time without saying "leopard." Furthermore, vervets will sometimes say "leopard" when there is no cat in sight, but when they are busy losing a fight with other vervets and want their opponents to scurry up the nearest tree, a diversionary tactic which can obviously be very helpful.

Among the vervets' other sounds is one which refers to eagles, one of their most alarming predators. But the vervets appear to be able to distinguish between the species of eagle which threatens them and others which do not, even though, seen from the ground, the various eagles are very similar.

It also seems to be the case that baby vervets improve the pronunciation of their vocabulary as they grow older. They "learn" to "say" things right. When an adult vervet hears the "leopard" call, it will check first to see who has made it before reacting – just in case it was a youngster trying to say something else.

A school of bottlenose dolphins: their communicative abilities may be very advanced, but relatively little is known about them.

Vervet monkeys discussing something, possibly how they feel about the photographer.

But a vocabulary of ten sounds, however sophisticated their use, is nothing compared with the variety of human sounds. The daily working vocabulary of a human is about 1,000 words which, crucially, are strung together in simple or complex sentences.

Robin Dunbar identifies a clear demarcation point between humans and other primates. Even the chimpanzees' quite detailed communication is not language: "Once you get into humans, you get this sort of burst of information-carrying capacity in communication."

Terrence Deacon has a robust view of the chimpanzees' abilities in this area: "I think we too often see evolution in hindsight and say, well, chimps are our closest relatives, it must be that they're closest to us in symbolic abilities.

"A good language experiment with animals is one that asks the question: what is the difference between teaching an animal by rote to use a series of gestures or symbols, to represent something, and to teach an animal to understand what it's communicating?"
(Interview, August 1993.)

In other words, you can teach a chimpanzee to look as though it understands the use of symbols: but can you teach it really to understand them, to fly free of the trainer?

E. S. Savage-Rumbaugh describes the activities of a pygmy chimpanzee named Kanzi in an article on language training of apes in the the *Cambridge Encyclopedia of Human Evolution* (Cambridge University Press, 1992.) Kanzi learned symbols and words by watching and listening. He understood a great deal of what was being said to him: "He soon became able to lead people to strawberries whenever they asked him to do so. He similarly learned the spoken names of many other foods that grew outdoors, such as wild grapes, honeysuckle, privet berries, blackberries and mushrooms, and could take people to any of these foods upon spoken request."

Kanzi was not able to produce speech, but he was able to respond to spoken requests such as 'Will you take some hamburger to Austin?' " By the time he was five, he was able to respond to sentences when he first heard them. He also became able not only to point to individual symbols but also to string sequences of symbols together which were "a primitive grammar and were not imitations of the experimenter."

The ability of chimpanzees to learn sounds and respond to them has led to their involvement, and that of other primates, in all kinds of human activities. A capuchin monkey (right) has been trained to help a disabled woman with many tasks including turning switches on and off, arranging food and drink and playing music tapes.

Savage-Rumbaugh concludes in her article that studies of ape language present a "serious challenge to the long-held view that only humans can talk and think. Certainly there is now no doubt that apes communicate in much more complex and abstract ways than dogs, cats and other familiar animals . . . Ape-language studies continue to reveal that apes are more like us than we ever imagined."

But in the minds of many scientists there remains a very clear distinction between what chimpanzees can do and what humans are capable of. Philip Lieberman, a major authority on language and communication, is unambiguous on this point. Chimpanzees cannot ever acquire human speech. They can be taught to recognize and act on a maximum of about 300 words. But a human child will know perhaps 5,000 words by the time it is four years old. Even if a baby chimpanzee is raised in a human household, exposed every day to human language, alongside human babies, it will never understand the syntax of even simple sentences that a three-year-old human child would grasp.

In spite of all the experiments and studies, and notwithstanding the emotional connection we have with them as our closest living relatives, there is an unbridgeable gulf between humans and chimpanzees. We can string sentences together, and they are not able to. We can speak about abstract things, and they cannot. We can think and talk about the future, about things which do not exist, and they cannot. Whatever the superficial similarities, evolution has taken us down a different road.

And it was a decisive turn into the new road. Nowhere in the world is there a simple language, one with a small vocabulary and just a few grammatical rules, suggesting an intermediate stage between grunts and calls – vocalizations – on the one hand and spoken language on the other. It appears that a species will go emphatically down one route or emphatically down the other. Language seems to be different in essence to other ways of communicating.

This idea is reinforced by the fact that human speech cannot be produced without a specific anatomy. The sounds we make demand a particular structure in our mouths and throats.

The larynx is much lower down in humans than it is in other primates. This means we can make much clearer, non-nasal sounds. With the larynx well away from the nose opening, we can make sounds which are loud and clear, especially the vowel sounds like "e" and "ou."

When human babies are born, their larynx sits high up in their mouths; at that stage of life, there is very little difference between the infants of the different primates. That early positioning of the larynx has a major advantage, as it enables the infant to drink and breathe at the same time. A human baby can suckle for lengthy periods without pausing and without

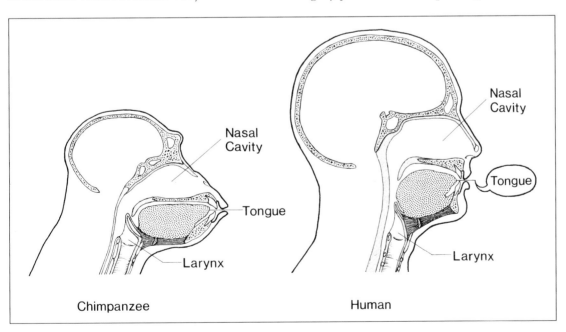

Nasal Cavity

Tongue

Larynx

Chimpanzee

Nasal Cavity

Tongue

Larynx

Human

The relative positions of the larynx in chimpanzees and humans. The adult human structure makes speech possible by the use of the upper part of the throat, the mouth and the nasal cavity – but at the risk of choking.

drowning – just like other animals. If you watch a dog drinking water, you see that it is capable of lapping away for far longer than any grown human could do.

As the human baby reaches about eighteen months of age, its larynx descends, enabling it to start making the right sounds for speech, but restricting its ability to eat and breathe simultaneously – at the same time creating an ever-present danger of choking. It is all too easy for humans to find pieces of food caught in the larynx itself. We are, as Philip Lieberman neatly puts it, "uniquely adapted to choking to death."

There are many examples in evolution of a vital adaptation taking place at a cost. Anatomically, evolution seems to create a series of compromises; advantages tempered by being achieved against the constraints of the existing raw material. Upright walking has the cost of disorders in the back and acute difficulties in childbirth. Speech has the cost of a structure of the larynx which is extremely dangerous. The fact of these compromises does of course emphasize the power of the need that the advantage be achieved. If it is so dangerous to speak, the pressure in its favor must have been very strong.

These adaptations of the mouth and throat, dangerous though they are to us humans, are simply not present in other primates. Chimpanzees, with their current anatomy, simply cannot speak.

The other indispensable adaptation, without which language would be impossible, is the appropriate change in the brain. This raises the question about the relationship between language, the development of the brain and the other physical adaptations to do with the mouth and larynx. Which came first? Unpicking the precise timing of the chicken and the egg here may be impossible: but the ability to deliver the complex language we use every day clearly hangs upon both a developed brain and the right physical equipment to make the sounds. And it may be that, as was the case with both the expanded brain and upright walking, once the process has begun, its continued development acts upon each element in the sequence of change: the brain drives speech, which drives the anatomy, which drives the brain, which drives speech, and so on.

We need to look more closely at the characteristics of our language. If we know what it does, we may learn why we needed language so much and when it first began.

One of the key features of spoken language is speed. Speech can compress individual sounds and deliver them in a very short space of time. The person hearing the sounds can then instantly decode them and understand the message.

Philip Lieberman says that mammals cannot normally distinguish between individual sounds if they are delivered faster than seven to nine sounds a second. Speech, in other words the manufacture of syllables, delivers sounds at a rate of ten to fifteen sounds a second, in a way that can readily be understood by the human ear.

If you say something very slowly, pausing between each word, the chances are that the person you are talking to will have forgotten the beginning of the sentence by the time you get to the end of it. This is why speed is so important to speech. The short-term memory in our brains has a very limited capacity, well suited to taking in rapid speech.

Speed, as Lieberman points out, obviously has practical advantages. The faster you can tell your friend that there is a leopard which he cannot see running towards him from behind the spiky bush on the left, the better. Examples of the advantages of speed are limitless. Lieberman suggests warnings, as above, but also reminds us about the benefits of quick explanations of tool-making techniques, the rules of football, or how to drive motor cars.

Speech squashes sounds into syllables, and it also squashes separate ideas into sentences through the use of grammar and syntax. A small child will say: "I saw the boy. The boy was big. The boy had a red hat." An adult crunches those thoughts together according to rules of language: "I saw the big boy with the red hat." And the same adult can make the sentence mean something slightly different in the same space of time: "I saw a big boy with the red hat." Saying that sentence out loud with slightly different emphases on each word can make

A uniquely human ability being demonstrated in a uniquely human institution. Traders in the Stock Exchange in Frankfurt, Germany use the ultimate level of rapid human speech to do their business and to create an exclusive bond with their fellow traders, probably the only other people who can understand them.

it mean other things again: did the big boy steal the red hat? What was he doing with a girl's hat, anyway? Why wasn't he wearing a red jacket? Why was he wearing a baby's bonnet?

Words have a set of literal meanings, but sentences and speech give them implications and a huge variety of initially concealed messages. In spoken language, the meaning of words only becomes apparent when they combine with other words in a certain way.

Philip Lieberman says that the key to the evolutionary importance of language lies in the fact that language and thought interact with one another. "It turns out," he says, "that general purpose areas of the brain like the pre-frontal cortex enter into speech motor control, syntax comprehension and, most importantly, thinking and problem-solving.

"Any development that would have enhanced language and pushed for enlargement in these areas of the brain would also enhance our thinking ability.

"It's a positive feedback process. You can't separate language, and the biological basis of language, from thinking." He goes even further, claiming that the parts of the brain which allowed us to speak also allowed us to think and act creatively.

Terrence Deacon makes the same connection. He says that the human ability to acquire a language relates directly to the biology of the brain. "The pre-frontal cortex is involved in analyzing combinations of things, sequences and dependencies among other things. And that's precisely the basis for how words mean what they mean." Sorting combinations of ideas is part of the same process as sorting meaning in combinations of words.

Language enables us to convey facts and ideas fast to other individuals and to large numbers of individuals simultaneously. It can disseminate ideas and information. What one person thinks can be in the minds of many other people in a matter of seconds.

In that sense it obviously has limitless practical uses in day-to-day life, especially day-to-day social life. One of the common threads in seeking explanations for evolutionary change is the demands of living in groups.

Language is a social glue. It creates and changes relationships, solves disputes, makes collective plans, and provides the means of education, of passing ideas and information on to the next generation.

Looking for a specific problem to which language might have been the solution, Terrence Deacon offers a fascinating idea which draws on both the importance of reproduction and the use of language to describe and capture symbolic as well as practical ideas. He admits that the idea is "way out on a limb," that it is a story rather than a scientific theory. But it is worth quoting at some length for its vivid evocation of life many hundreds of thousands of years ago, when so many things we take for granted today were simply not present.

"Let me lay out the problem that I think language is in part the solution to, or rather that symbolic reference is the solution to." The problem is one in which hominid males are gathering meat. They are gathering food for themselves, and, crucially, for their offspring – and the children are not with them, because near the sources of meat are likely to be all kinds of other, dangerous scavengers. So the meat is to be collected and taken back to a safe place to be eaten.

It is important to the males that the offspring they feed are their own. In much of the rest of the animal world, this problem is overcome by isolating breeding pairs into their own territory. But in the early hominid world, "you have a group of primates who have to co-operate to get food. You get a group that's forced to live together and work together, of males and females, all of reproductive age – and nevertheless you need to guarantee to some extent, in a genetic sense, that you know you're giving food to offspring of yours.

"I think the problem that's posed by moving into an environment where meat becomes a necessary commodity for raising children, is that you find ways to build in predictable social

A Catholic wedding at Magdalen College Chapel in Oxford University: communication of an important statement through ritual, movement, and sound.

behavior, and specifically around sexuality – socially agreed upon inclusion and exclusion of sexual relationships.

"These kinds of relationships are not just mating relationships, they are, in a sense, promises. They are communications about a possible future, about what should and should not happen. This is not something that you can represent with a call or a gesture.

"I think the first context in which symbolic representation evolves is something like a marriage ritual, a social, public determination of sexual obligations and reproductive exclusions."

This idea cannot be either proved or disproved. The fossil record offers evidence about brain size, which provides pointers as to when language may have begun, but no fossil can explain the exact circumstances in which it began. Modern evolutionary science builds on the fossil record through the study of living humans and living relatives of humans. Ideas about the past are put together by an understanding of the present. The attraction of Deacon's notion about the origin of language is that it draws together these various strands of thought about evolutionary change, and that it hangs upon an understanding of language as being to do with ideas, with thinking, as well as with information.

Robin Dunbar draws the origin of language back firmly to the pressures of living in groups. The larger the group of individuals, the greater the need to find ways which are not too time-consuming of operating all the relationships necessary for the smooth running of the group. Living in the open involves spending an enormous amount of time finding food; the amount of time other primates spend cementing their relationships by grooming one another places a glass ceiling on the size of the group. "What language seems to do," says Dunbar, "is to allow you to overcome that barrier and use what time you have in a much more efficient way. You can talk to several different people at the same time. You can talk while you eat, talk while you walk or talk while you work. This allows you to expand the network of individuals whom you have a relationship with.

"Some ecological pressure demands bigger group sizes, so brain size has to increase to make that possible, until you hit this Rubicon at which you cannot increase group size any further – the animals don't have enough time to bond in their relationships.

"At that point language appears, to make possible these super-big groups in humans. You have a sequence of stages in which ecology determines group size, which determines brain size, which finally determines language."
(Interview, September 1993.)

The obvious next stage, which flows from this analysis of the process by which language began, is to put dates to the ideas.

There are two broad conflicting views about this. There are those who put a very early date on origin of language, broadly to about two million years ago, the time when the fossil record shows that the brain had begun to enlarge. And there are those who propose a much later date, broadly coinciding with the time, less than 100,000 years ago, when symbolic expression in the form of art and sculpture began.

In 1959 Mary Leakey and her son Jonathan found some fossil remains in the Olduvai Gorge in Tanzania which seemed to be markedly different from everything else known at that point. They dated to around 1.8 million years ago, and the size of the brain case was significantly larger than those of the *Australopithecus afarensis* and *A. africanus*. There were other differences as well (the teeth were smaller and narrower), but the greatest interest focused on the brain.

Phillip Tobias and Louis Leakey concluded that, in proportion to body size, the new specimen's brain was 50 percent larger than that of the *Australopithecus africanus*. More specimens came to light in the same area, and a case gradually began to be built that this was a new species of hominid. It would sit somewhere between the australopithecines and *Homo erectus*, widely accepted at the time as the first species of *Homo*, dating back to around one and three quarter million years ago.

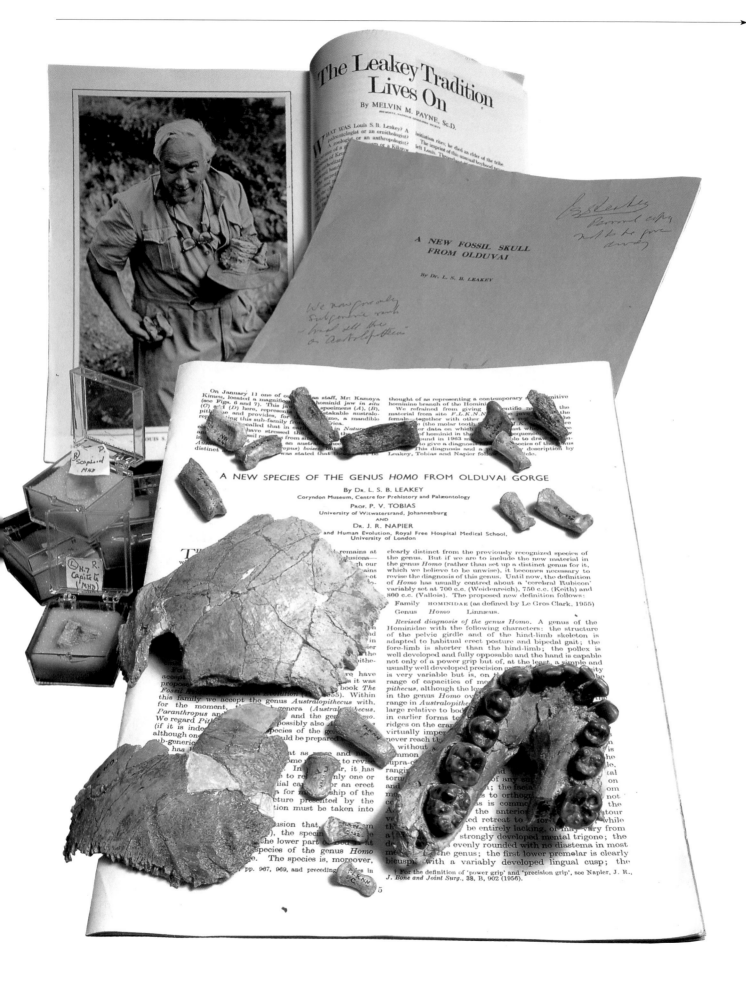

The Leakey Tradition
Lives On

By MELVIN M. PAYNE, Sc.D.

A NEW FOSSIL SKULL
FROM OLDUVAI

By Dr. L. S. B. LEAKEY

A NEW SPECIES OF THE GENUS *HOMO* FROM OLDUVAI GORGE

By Dr. L. S. B. LEAKEY
Coryndon Museum, Centre for Prehistory and Palæontology

Prof. P. V. TOBIAS
University of Witwatersrand, Johannesburg

AND

Dr. J. R. NAPIER
and Human Evolution, Royal Free Hospital Medical School,
University of London

OPPOSITE: *Louis Leakey with the original specimen of* Homo habilis, *discovered in 1960 in the Olduvai Gorge in Tanzania.*

The Olduvai Gorge (left) is cut into the southeastern Serengeti Plains in Tanzania. Large numbers of early stone tools and of fossil remains have been found there, including the original Homo habilis *and "Zinj." KNM-ER 1470 (below) was found much further north, at East Turkana in Kenya. It was reconstructed from 150 fossil fragments and was originally dated at over 2.5 million years old, although this date was later revised to less than 1 million years.*

In 1964 a paper appeared in the journal *Nature*, under the joint names of Phillip Tobias, Louis Leakey and John Napier, launching a new species called *Homo habilis* – a name dreamed up by Raymond Dart to convey the idea that this was the first species of possible human ancestors to be clever enough with its hands to make and use tools.

The area around the Olduvai Gorge had long been known to be littered with stone tools. They were being found there as early as the 1920s, providing tantalizing signs that someone had once inhabited the area who had routinely shaped useful implements out of the natural geology of the region. But it had taken forty years or so to arrive at some conclusions about who these ancient inhabitants had been.

The best known of all the *Homo habilis* fossils, found by Richard Leakey on the east side of Lake Turkana in Kenya, is known as KNM-ER 1470, from its catalogue number in Kenya's national museum. It was discovered at Koobi Fora in 1972; it had a large brain, with a long, broad face. None of its teeth were preserved, but the sockets for them are large by modern human standards. Some of the skull's features, such as the cheek bones, are similar to those of the australopithecines. But the brain-case is markedly more human-like in its shape.

The brain size of the new species was sufficiently enlarged, and its shape sufficiently visible from casts made of the insides of the skulls, for Tobias to conclude that *Homo habilis* had the neurological basis of speech. This was, and is, an extremely dramatic and controversial claim.

"The way brain evolution works is that function and structure go hand in glove with each other. I believe that not only has this human burst of brain enlargement come on the scene with *Homo habilis*, but an entirely new dimension, spoken language, has appeared for the first time." (Interview, August 1993.)

Terrence Deacon comes to much the same conclusion. From two million years ago, with *Homo habilis*, until about 500,000 years ago, with *H. erectus*, changes in brain structure had been taking place which resulted in brains very similar to ours today. Deacon believes that

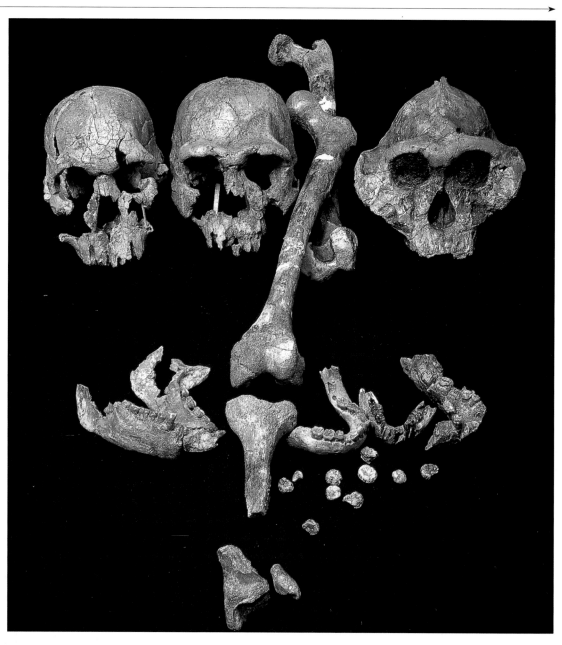

A collection of hominid fossil skulls and other bones found at East Turkana in Kenya.

the changes were driven by language, and that the process began right at the beginning of the period probably occupied by *H. habilis*, over two million years ago. The changes coincided, he says, with movements of climate, the appearance of stone tools, and dramatic changes in the social lives of the hominids.

In the time of the australopithecines, there was considerable difference between the body sizes of males and females, suggesting arrangements in which individual males were surrounded by something resembling harems of females. By the time of *Homo erectus*, body size between male and female had significantly evened out, suggesting patterns more in the line with human behavior today.

"At this transition at about two million years, a whole series of things from social life, ecology, ways of making a living, ways of getting food, to the structure of the brain, all seemed to change at once. The species went through a sort of systematic restructuring of the way of making life."

(Interview, August 1993.)

Deacon does not suggest that *Homo habilis* was speaking the way people do today. He thinks language then could have been a combination of speech and signs: "It might have

---

looked a lot like a modern ceremony or ritual, with sounds and movements and exchanged objects all involved in the communication."

But the key is that it would have included the ability to refer to the abstract, to matters other than the here and now. For Deacon, the development of the brain is caused by language which in turn is caused by symbolic communication. "Even that first language-like communication had to have the ability to refer to things in the future, to refer to the possible, to make promises, to anticipate. The first language, in that sense, had the power that modern languages have, the power of the abstract."

When Louis Leakey and his colleagues identified *Homo habilis*, the new species became a kind of emblem of the ancient lineage of humanity. For Tobias, it was the point at which a line was crossed between "ape-ness" and "human-ness;" the australopithecines were hominids whose characteristics were predominantly ape-like, whereas *H. habilis, H. erectus* and *H. sapiens* were recognizably human.

Tobias says that Louis Leakey had been convinced since the 1930s that *Homo* would turn out to go very far back in time. "He was exultant that his early idea of the high antiquity of the genus *Homo* seemed to be now confirmed by this creature.

"We can say that somewhere between two and a half and two million years ago, a new dimension in human evolution appeared on this planet. Instead of the hominids behaving like animals, like chimpanzees and gorillas, they started behaving in an extraordinary new fashion. Brains dominated the scene now; what you could do with your brains and culture came to be a predominant tool kit for dealing with the ravages of the environment." (Interview, August 1993.)

Talking

The difference in size between males and females is much greater in early hominids than in modern humans, and this may in turn reflect a difference in social systems. For the australopithecines, the system may have been like that of chimpanzees, with a core of related males defending a territory, and sexually mature females migrating in from other groups. The males compete with each other for access to the females, but do not form long-term bonds with them. Human social systems vary enormously, but they are characterized by long-term bonding between males and females, and less overt sexual competition between males.

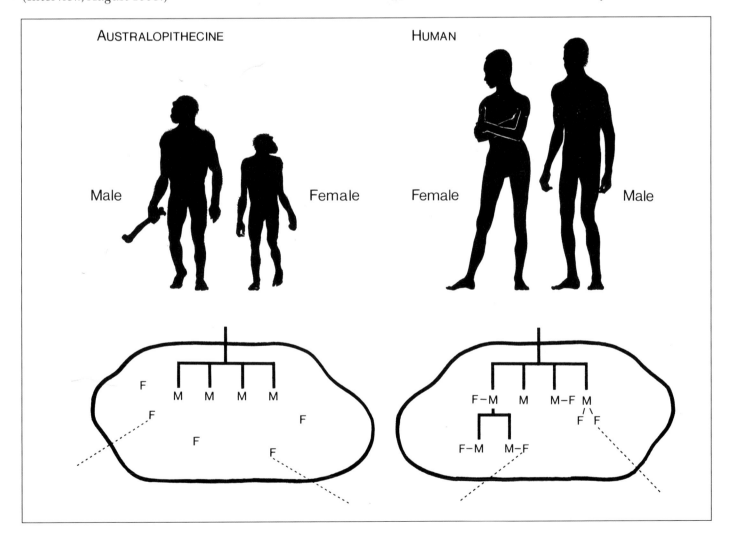

Homo erectus, *which appeared around 1.75 million years ago, was probably the first species of hominid to control and use fire – and also the first to move outside Africa. It is likely that the spreading population reached parts of Asia very early during the existence of the species. These new forms of behavior plainly suggest greater communicative skills.*

Leakey had found the turning-point in this scenario. The big brain and the use of tools, the capacity for language – these were all very human attributes. But the problem with this argument is that even if it is the case that *Homo habilis* possessed these attributes, there is very little sign in the archaeological record of anything much being done with them. As far as can be determined, the style of stone tools did not change for over a million years, and there is no sign of *H. habilis* controlling fire or grappling with any other technological change.

Not until *Homo erectus* comes on the scene does new evidence of major changes in behavior and technology begin to appear. The weakness of reading too much into the brain size of *H. habilis* lies in the absence of any indication of consequent change. It is not at all clear that he did in fact begin to behave in an extraordinary new fashion.

Hunting wild animals is one of those activities which appears in most lists of specifically human behavior. Hunting implies social organization and planning; it suggests bands of people catching food and bringing it back to a home base. There was an obvious temptation to associate the stone tools and fossil remains in the Olduvai Gorge with the systematic catching and killing of animals rather than with scavenging on the bodies of animals which had died for other reasons.

The more of these human-like attributes that can be attached to a particular fossil, the greater the star quality of the fossil. The combination of great age and human qualities is the trump card in the fossil-hunting game.

In the 1960s, when *Homo habilis* was being named, the hominid lineage was thought to be older than it is today. The molecular studies which have dated the common ancestor with the apes at five to eight million years ago have only been gaining authority in the past twenty years or so. Before then, it was believed that distinctly hominid species existed very much earlier.

A fossil discovered in India in 1934 by an American geologist, G. Edward Lewis, was given the name *Ramapithecus*. It was believed through the 1960s that it was an ancestor of the australopithecines and was a very early hominid, dating to about fourteen million years ago.

Conclusions reached about a particular fossil must always be examined in the light of prevailing knowledge about human evolution at the time of the discovery. So in the context of a lineage of such antiquity, it is perhaps not so surprising that *Homo habilis*, dated to only two million years ago, might be seen to have developed a very wide range of human characteristics.

Today's contextual knowledge is different. The ape–hominid split is known to be much more recent, and the process of evolution was not one in which each new fossil find has a slot to fit into on a progressive ladder from ape to human. The multiple hominid species of the past are all seen in their own right as having come into existence as a result of interaction with the climate, with the environment and with other competitive species.

Plainly, humans have an ancestral line leading back through time. But the purpose of palaeoanthropology seems to have changed. It no longer stands or falls on nailing a particular fossil to the human lineage, but on understanding the process of change, understanding what brought a species into existence and what, ultimately, caused its extinction.

This means that the attributes of each species can be accepted for what they are rather than having a value attached to them according to the species' proximity to modern human behavior.

This in turn means that understanding an early species stems not from comparing its various characteristics to those of today's *Homo sapiens*, but from exploring the circumstances of the world in which the early species actually lived. And this is highly relevant to the timing of the beginning of language.

Lewis Binford, of the Southern Methodist University in Dallas, has been extensively involved in excavations in Tanzania and has carefully examined the inferences which have been drawn about *Homo habilis*.

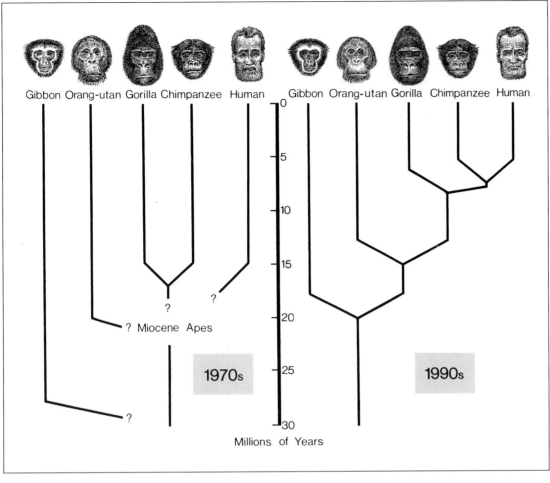

Gibbon   Orang-utan   Gorilla   Chimpanzee   Human   Gibbon   Orang-utan   Gorilla   Chimpanzee   Human

? Miocene Apes

1970s

1990s

0
5
10
15
20
25
30

Millions of Years

Until the 1970s, prevailing opinion held that humans and apes had separate evolutionary histories going back as far as 15 million years or more. The point at which the gibbons split off from the evolutionary line was put as far back as 30 million years or more. But in the 1980s these views were overturned, especially in the light of molecular and genetic research. Today it seems that all living apes have evolved within the last 20 million years or so, with humans and chimpanzees sharing a common ancestor as recently 7 million years ago.

"In a technical sense we as archaeologists don't have any facts. In science a fact is defined as the property or characteristic of an event. And the events that we are interested in, we can't observe."

(Interview, August 1993.)

As an example, he quotes the juxtaposition in the Olduvai Gorge of fossil animal bones and stone tools. The connection between them is often seen as a fact, in the form of a hominid agent. The hominids had been hunting, and had brought sections of their catch back to a home base to feed the family.

But for Binford, the connection between the bones and the stones can never be anything more than inference. "When they found the stone tools with the bones, the inference was made quickly that they were already hunting and therefore already much like us.

"A number of people have argued that it was the shift to hunting which had made us human, because, they argued, you had to co-operate. But the work I had done with ethnographic peoples in Africa, particularly in the Kalahari region, where they regularly scavenge off lion kills, suggested that there was another possibility. That they weren't killing many of these animals, but were exploiting the entropy of a very complex ecosystem through scavenging and stealing."

The picture Binford builds of life at the time of *Homo habilis* involves a much more mobile and opportunistic strategy than that put forward by those who find human characteristics in the hominids. "The feeding was much more like the rest of the apes. They encounter food, they eat it, and then when they begin to get short they move on, and they get some more, and they eat it there."

His arguments apply not only to feeding, but crucially to our dominant theme – communication. "I have no doubt that the early hominids had very high-level communication systems, but I suspect they didn't have language." Returning to the model of the computer, Binford agrees that the hardware, in the sense of the enlarged brain, was in place at the time of *Homo habilis.* But their software was a very early version.

"I don't think these guys were dumb, I think they had information-processing, communications systems of some elaboration. But the quality of language which makes it very different is the ability to abstract, to anticipate events not experienced. I don't think animals with the finest communications systems do that – so there's a qualitative difference." The new, upgraded software for language, in the Binford metaphor, was not available until very much more recently, within at the most the last 100,000 years.

Not until then, he points out, does the archaeological record offer evidence of ornamentation, painting and other visible symbols, all of which go alongside the concept of language. "Art itself is graphics, and you can't assign meaning to a little scratch on the wall unless you've got language." So it is reasonable to suppose, he thinks, that the great explosion of art and sculpture which is visible from about 40,000 years ago is associated with the emergence of language as distinct from earlier forms of communication.

Philip Lieberman takes a similar view, placing the origin of language, as we speak and understand it, at about 100,000 years ago. He thinks the earliest australopithecines would have had, perhaps, somewhat better communicative abilities than modern chimpanzees, and that language ability would have progressed in fits and starts with each new hominid species.

Robert Foley says that a possible solution to the conundrum, which would take emphasis away from having to tie language to the big brain, is to separate language from thought. Just because the chimpanzees' calls and gestures are mostly practical in their meaning does not necessarily imply that they are not having abstract thoughts. The research by Robert Seyfarth

Lewis Binford points to the behavior of modern communities in the Kalahari in southern Africa to support his view that early hominids may well have survived very successfully for many hundreds of thousands of years without becoming "hunters."

There is no doubt that by the time paintings like this were being made, in the Lascaux caves of south-western France during the last Ice Age, the use of language was fully developed.

and Dorothy Cheney suggests that monkeys think more than they communicate about those thoughts; the need to think may be quite separate from the need to communicate the outcome of your thinking.

If that were the case, and if the principle applies also to evolution, then the early hominids, including *Homo habilis*, might have been thinking complicated and abstract thoughts, and may have been communicating them in some way; and it may not have been until much later that the social and other pressures of life brought modern language to *H. sapiens*.

The mental capabilities associated with language may have begun as long ago as two million years. There is virtually no doubt that with all its extraordinary properties of speed, volume and abstraction, language was effectively at the level that it is now by about 100,000 years ago.

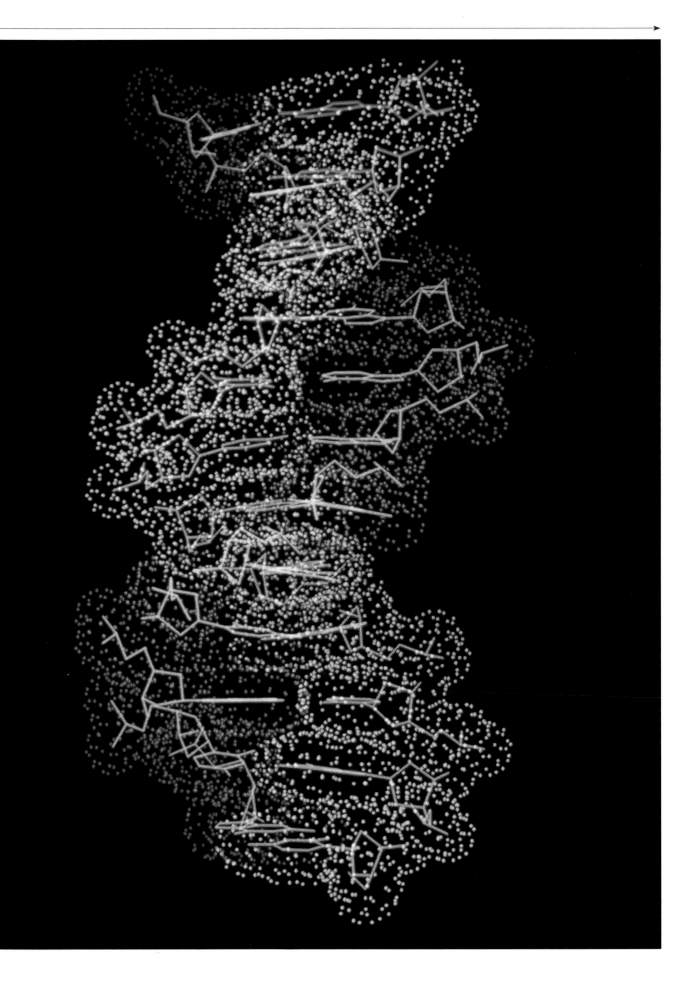

5 Colonizing

A major preoccupation of evolutionary scientists has been the question of how humans got to be where they are today geographically. Every square mile of the habitable world is inhabited by *Homo sapiens* – we are the only animal species which has achieved virtually complete coverage of the surface of the world.

This is a mark of biological success. Answering questions about how it came about also answers questions about who we are and where we came from in the first place.

It is an emotional problem as well as a historical one. Every human, apart from identical twins, is different in some degree from all others. More significant to some people than the minor individual differences of height, size and appearance are the distinctions which separate races from one another – people who live in different parts of the world and who look different.

Most biologists would not even use the word race. In the context of human diversity, it has no scientific meaning at all. In fact you can find 80 percent of the world's human genetic variation in any single population. Biologically, all humans are in the same species, and they can and do mingle and breed with one another freely, wherever they come from and whatever they look like. In today's fluid world, with rapid and widespread travel and communication, there is no practical possibility of a particular human population becoming so isolated that a genetic barrier would exist between it and other humans. Even though there are still many parts of the world where individuals may never travel from the immediate surroundings of their home base, this stability and relative isolation are more than offset by the tendency of people from other places to travel to them.

That said, there are obvious physical differences between the peoples of the world. Some are black, some are white, some are yellow, and with these skin colors go a set of characteristic hair and facial features. For some reason humans like to endow these differences with great significance, to attach ideas of superiority to one appearance over another. This habit has led to great conflict, pain and destruction. Racism is one of the most negative manifestations of the working of the human brain.

Perhaps the most important single contribution that the study of human evolution can make, especially to the non-scientific community, is to set these racial differences in their his-

An image of DNA, realized through computer graphics. DNA consists of two linked strands of nucleotides coiled into a helix; the way the strands are arranged, and the way the bases which make them up are paired, determines the genetic information passed on to the next generation. DNA studies have cast a whole new light on human evolution.

109

Phillip Tobias looking over the site of the excavations at Sterkfontein in South Africa, the scene of many vital discoveries.

torical context. People may still feel hostile towards one another for all kinds of irrational reasons. But the study of evolution can demonstrate just how misplaced some of those reasons are.

It is hard to imagine how difficult it must have been to be involved in the study of human origins in South Africa during the worst years of apartheid. Not only was the academic community in that country effectively cut off from the rest of the world by a political gulf and by active boycott; but also the very subject of palaeoanthropology went to the core of the country's political system, because its overriding message is the lack of significant differences between black and white populations.

Phillip Tobias held the Chair of Anatomy at Witwatersrand University through many of the darkest years and he has spoken out unwaveringly for human rights. He says that part of the reason we like to study our past is simple curiosity, a feature we share with all other mammals. "But there are other messages that we may get from the past. I suppose it would be true to say, when I show you some ape-men from Sterkfontein, that those are the great grandparents, not just of Africans but of everybody on the face of the earth. If you want an argument in favor of the brotherhood of man, I know no more telling argument than that we all share a common ancestry.

"That to me is a scientific underpinning of the ethical principle of brotherhood of humanity, if ever one needed such an underpinning."
(Interview, August 1993.)

Tobias's principle of brotherhood comes from the knowledge that wherever humans are today, they share a distant ancestor who lived in Africa. We know with virtual certainty that we are all descended from an ape-like creature, which is also the ancestor of today's chimpanzees, and that the ancestor lived around five to seven million years ago. This knowledge comes not from the fossil record, but from molecular biology: the comparative study of the DNA codes of today's humans and today's chimpanzees.

Much of the pioneering work in this field was done by the American scientist Morris Goodman, currently at Wayne State University in Ohio. Prior to the late 1950s it was generally believed that hominids had been evolving on a line of development separate to that of all the Old World apes – chimpanzees, gorillas, orang-utans and gibbons. Conventional wisdom held that a common ancestor between the great apes and the hominids preceded the division of the apes into different species. This allowed, as we have said before, an ancient lineage for humanity and maintained a comforting conceptual barrier between the idea of being an ape and the idea of being a human, despite a common ancestor in the dim and distant past.

Morris Goodman's protein studies of living species led him in the late 1950s to propose a radical rethink of these relationships: "It was quite clear to me that the chimp and gorilla were much more closely related to humans than they were to the other great ape, the Asian orang-utan, and still more so than to the gibbons."
(Interview, August 1993.)

These studies suggested that the divisions between chimps, gorillas and humans were more or less equal; but that together, the three species were significantly separate from the other apes.

Goodman is a modest man, but he admits under pressure that his and his colleagues' studies were the first time that molecular biology had had a major impact on our understanding of who humans are most closely related to. There was some earlier published work on the subject, in 1904, but it was largely ignored. The idea that the hominids split off from the rest of the ape family in the relatively recent past obviously presented a head-on chal-

DNA studies reveal that orangutans are much less closely related to humans than chimpanzees and gorillas are.

lenge to the notion of a hominid lineage of such great antiquity as to separate human life comfortably from that of all the apes.

Goodman went further. Comparing the DNA structure of chimps, humans, gorillas and the other apes, he has found that there is only a 1.7 percent difference between chimps and humans, and a 1.9 percent difference between a human and a gorilla, or between a chimp and a gorilla. The difference between all three and an orang-utan is about 3.7 percent.

DNA is made up of strings of four bases, represented by the letters A, C, G, and T. These strings wrap around one another in the famous double helix; each string complements another – the A is always laid on a T, and a T on an A and a G on a C and a C on a G.

For DNA to be passed on to the next generation, it has to make copies of itself. The helix unwinds, and each string is used as a template to create another double helix, consisting of the same sequence of letters laid on each other. The copies, in cells associated with reproduction, are what makes the next generation. The copies are the threads which link generations to one another.

The permutations of the sequences of letter create our various traits. The sequences make up genes, which code for amino-acids, which are built into proteins – which are the building-blocks out of which our bodies are made.

Sometimes, however, the copies are inexact, errors are made; an A might turn into a G or a C into a T. This is called a mutation.

If there were never any changes, there would be no evolution. If the change is harmful, it will not spread through the population. But if the change is beneficial, and gives the organism an advantage in its environment, then it will spread, and evolution will have occurred. These positive changes will affect the morphology – the physical attributes and appearance – and the biochemistry of a living creature, to make it more suited to successful survival in a given environment.

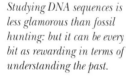

Studying DNA sequences is less glamorous than fossil hunting: but it can be every bit as rewarding in terms of understanding the past.

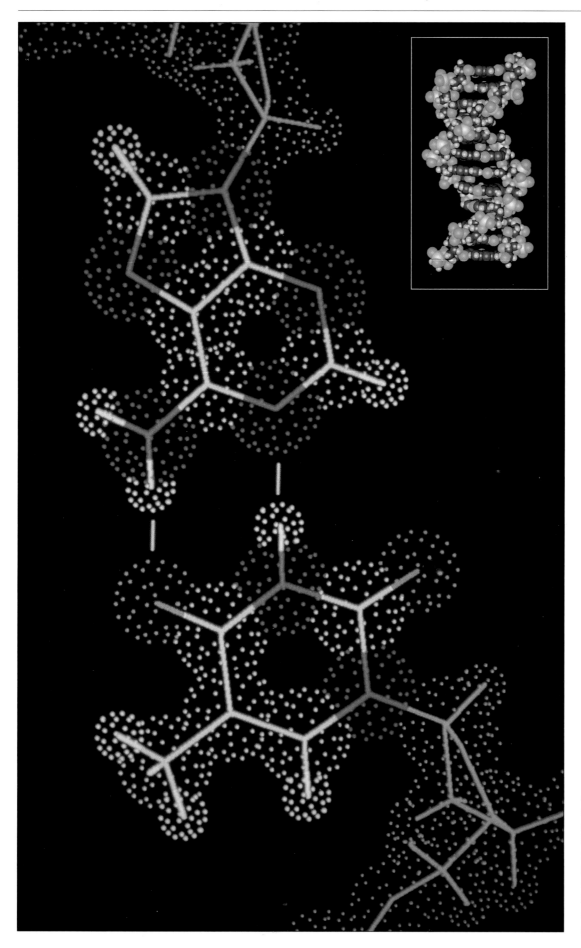

A computer graphic illustration of the linking of a base pair in a DNA strand. The adenine base (A) is at the top, linked to thymine (T) by two hydrogen bonds (pink, center). Inset is a representation of a section of a DNA molecule, the genetic material of most living organisms.

Nuclear & Mitochondrial DNA

DNA is a thread-like molecule, sometimes many centimeters long. In humans, over 99.99 percent of a cell's DNA is packed into the cell nucleus, with a small amount occurring outside the nucleus in small structures called mitochondria. Nearly all our genetic information is carried by DNA inside cell nuclei. Only 2 percent of nuclear DNA actually seems to encode information for making proteins. The function of the rest, known as "junk DNA," is still unknown.

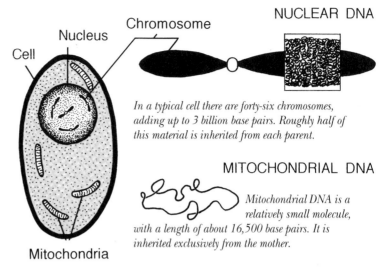

NUCLEAR DNA

Cell
Nucleus
Chromosome

In a typical cell there are forty-six chromosomes, adding up to 3 billion base pairs. Roughly half of this material is inherited from each parent.

MITOCHONDRIAL DNA

Mitochondrial DNA is a relatively small molecule, with a length of about 16,500 base pairs. It is inherited exclusively from the mother.

Mitochondria

DNA structure & replication

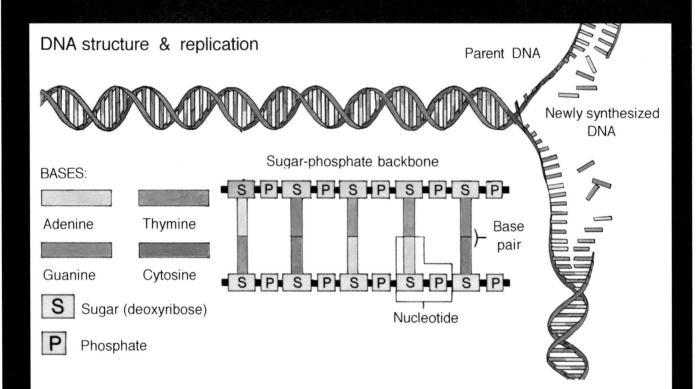

Parent DNA

Newly synthesized DNA

Sugar-phosphate backbone

BASES:

Adenine   Thymine

Guanine   Cytosine

S Sugar (deoxyribose)

P Phosphate

S P S P S P S P S P

Base pair

Nucleotide

DNA can be visualized as a twisted ladder, with ten rungs for each 360-degree twist. The sides of the ladder are a uniform backbone made up of repeating sugar and phosphate groups; the variable part of DNA is its sequence of bases, which pair up to form the rungs. The basic unit of DNA is a nucleotide, which is a sugar-phosphate link plus an attached base – each half of the DNA ladder is a linked series of nucleotides. The ordering of bases along the series or chain varies, and this order carries the genetic information. When DNA replicates, the two chains unwind and separate, like a zipper opening. Each parted chain then acts as a template for the formation of a new, complementary chain.

DNA finger printing can now be used as a highly reliable means of identifying individuals or of establishing whether people are related to one another.

Most changes in the DNA chains, however, are neither harmful nor beneficial. They are random, neutral variations in the pattern, which have no apparent effect on organisms, but are the most useful to enable us to follow the relationships between populations and species. Goodman explains: "These changes in the DNA caused by mutations, and in particular those that spread through the population and then the species as a whole, are what allow us to determine the kinship relationships of one species to another. The more closely related the species are, the fewer the differences. This is easily determined by working with the DNA itself."

The DNA can be a much more reliable guide in assessing relationships than mere appearance. The difference in appearance between humans and chimps is great or small depending on your angle of perspective. Goodman gives another illustration: domesticated dogs only appear about 10,000 years ago – yet to look at a Great Dane by comparison with a Chihuahua, a casual observer would assume that vast genetic difference accumulated over a much longer period. In truth, the genetic difference is very slight and recently acquired.

Thus DNA studies are more fundamental to the process of evolution than mere observation, and they refine the earlier evidence from proteins. They confirm that there is a greater gap between humans and orang-utans than between humans and chimps. So it follows that the split between orang-utans on the one hand and humans, chimps and gorillas on the other must have predated the splits between chimps, humans and gorillas. And, most provocatively, it looks as though the split between humans and chimps came later than that between chimps and gorillas.

Even more surprisingly, dates can be attached to these evolutionary splits by working on the basis of some degree of consistency of genetic change, relating it to the fossil record. Where the sequences of bases have no effect – in what is called junk or non-coding DNA – the difference between species will accumulate at a constant speed, simply as a function of the rate of mutation. This is known as the molecular clock. Like any clock, it sometimes goes too fast or too slow, or even stops. But in general it gives a good estimate of when things have happened in evolution.

The molecular clock can be calibrated from a known date in the fossil record. Palaeontologists reckon from the record now that the split between the orang-utans and the other species occurred about sixteen million years ago. If that represents a DNA divergence of 3.6 percent, it follows that a DNA divergence of 1.7 percent must have occurred about half that length of time ago.

This is, in very simple terms, how Goodman arrived at the figure of around eight million years ago for the split between humans and chimps as against gorillas, and the date of around seven million years ago for the last common ancestor of the humans and chimps.

There is no fossil record for the common ancestor population. Such a find is one of the great Holy Grails of palaeoanthropology. The earliest current fossil, subsequent to the common ancestor, is *Australopithecus afarensis* – Lucy – and the footprints made by members of that species at Laetoli in Tanzania around three and a half to three and three-quarter million years ago. The refined character of the footprints, and of the *A. afarensis* fossil bones, suggest that upright walking was well established by that time, which leads to an estimated start date for Lucy's species of around four and a half million years ago, which of course is a near match of the date offered by Goodman for the common ancestor.

There is huge significance in this date, greater even than the debate around the evolutionary causes of upright walking. Because it means that the oldest recorded hominid fossil is located in Africa. Nothing found so far anywhere else in the world is that old.

A Dutch anatomist, Eugene Dubois, found fossil remains in Java in the 1890s, which he believed to be very ancient "missing link" remains. Today they are dated at no older than one million years, although some very recent studies place some of the Java remains at over one and three quarter million years ago. A similar date is attached to the many fossil remains discovered since the 1920s at the Zhoukoudian cave site in China ("Pekin Man.") At the time, these discoveries were thought to be of great importance because they lent weight to the idea of an Asian origin for humanity. But it is now clear that they are over two million years younger than the oldest African fossils.

Eugène Dubois (center, standing) with Sir Arthur Keith (top left) in Cambridge in 1898, five years after Dubois had found Java man. Sir Arthur Keith refused to accept Dubois' assertion that his find represented a link between ape and man, and out of pique Dubois is said to have hidden the bones under his dining room floorboards.

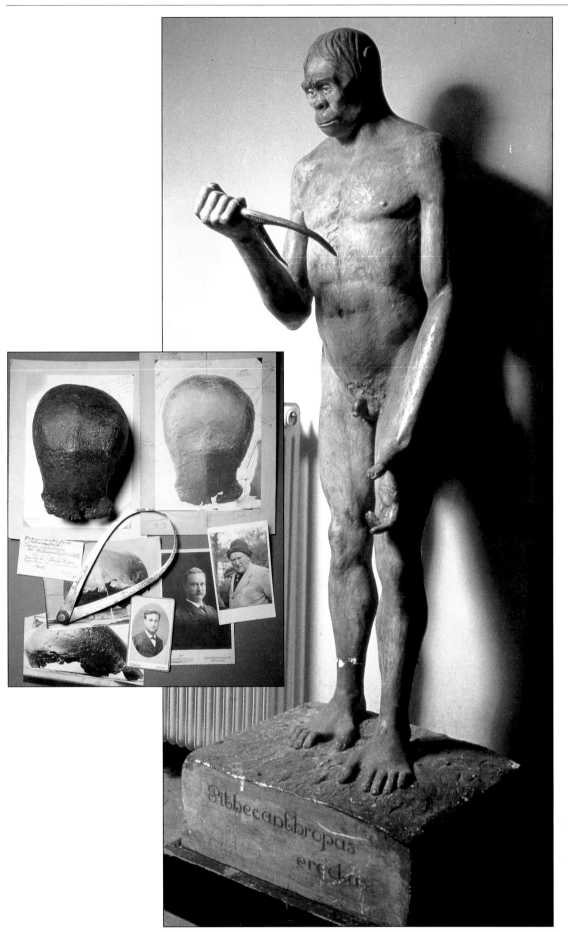

A full-size reconstruction of Java man, made by Eugène Dubois himself. The remains on which he based the model, found at a bend on the Solo River in central Java, Indonesia, included a complete thigh bone and skull cap (inset).

Ape Man

A plaster cast reconstruction of Pekin man (above), and the skeleton of Turkana boy (below).

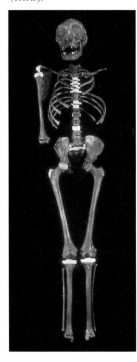

A classic australopithecine skull, compared with the much more modern-looking Homo erectus, *represented by Java man, and Pekin man (top).*

The evidence of an African origin for all hominids piles up very impressively. There is the age of the African fossil finds and their consistency and quantity. There is the fact that Africa is the home of the hominids' closest living relatives, the gorillas and chimpanzees. There is the cumulative evidence that Africa was the site of truly extraordinary ecological and climatic changes at the time we know that evolutionary change was taking place. In spite of the prejudice which resisted the idea for many decades, there is now no serious scientific doubt that Africa was the starting point, the home of the common ancestor of hominids, chimpanzees and gorillas.

But there have been many hominid species, of which *Homo sapiens* is only the most recent. The fact that the hominid line as a whole began in Africa does not necessarily mean that the ancestors of today's humans came from there. The age of the species *H. sapiens* is not certain, and sometimes it is broken down into two stages: archaic *H. sapiens* and then *H. sapiens* itself.

The age of fossils associated with archaic *Homo sapiens* is normally put at between 200,000 and 500,000 years ago, and of fossils described as modern *H. sapiens* at about 120,000 years ago. In either event, the species is plainly very young, set against a hominid time-scale reaching back about seven million years: so where and how did this new species originate?

The first hominid species assigned to the genus *Homo* was *H. habilis*. We know that *H. habilis* lived in Africa from somewhat over two million years ago, that they had relatively large brains as compared with the australopithecines, and that they used stone tools. They were meat-eaters, but there is no evidence that they used fire, or that their population spread much beyond eastern and southern Africa, the sites where their fossil remains have so far been found.

A further species of *Homo* appears in the fossil record from about one and three quarter million years ago – *H. erectus*, whose span appears to cover over one million years. They had significantly larger brains than *H. habilis* and flatter, more human-like faces. Some of the best *H. erectus* finds have come from Northern Kenya, including the nearly complete skeleton of a boy aged about eleven years from a site called Nariokotome on Lake Turkana.

Turkana boy was found in 1984. He was not yet fully grown, but was already about five foot three inches tall, with long legs and a heavy build. He is a very interesting individual, not least because he is much larger than a similarly aged modern young adolescent would be. His age is established from his teeth – not all his permanent teeth had appeared and he still had some deciduous baby teeth.

Holly Smith, at the University of Michigan, has made a special study of Turkana boy, and her conclusions help to place *Homo erectus* in relation to the other *Homo* species. "If you look at his teeth in detail and do the finest possible assessment of his dental age, and if you assess him as though he were a human child, you get a suggested age of about eleven years.

"The problem is, if you look at his skeleton, he's very tall for an eleven-year-old. His skeleton has the maturation you might expect of a child of thirteen, and it has the size of a fifteen-year-old – so we have some discrepancy here."

Her conclusion about how to resolve the discrepancy relates to the way life cycles in general have evolved alongside other, more obvious evolutionary changes. Chimpanzees grow up much more rapidly than humans. They are finding their own food by the age of four and are fully adult by the age of about twelve years. Modern humans develop much more slowly (and live longer.) It is estimated that human babies take about a year to reach the range of capabilities of a chimpanzee at birth.

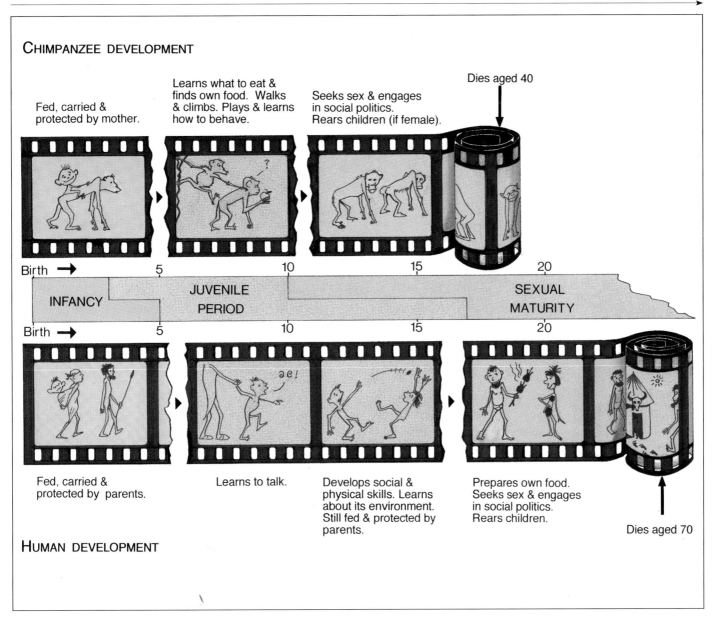

CHIMPANZEE DEVELOPMENT

Fed, carried & protected by mother.

Learns what to eat & finds own food. Walks & climbs. Plays & learns how to behave.

Seeks sex & engages in social politics. Rears children (if female).

Dies aged 40

Birth →   5   10   15   20

INFANCY   JUVENILE PERIOD   SEXUAL MATURITY

Birth →   5   10   15   20

Fed, carried & protected by parents.

Learns to talk.

Develops social & physical skills. Learns about its environment. Still fed & protected by parents.

Prepares own food. Seeks sex & engages in social politics. Rears children.

Dies aged 70

HUMAN DEVELOPMENT

In Holly Smith's view, the early australopithecines will have had life cycles similar to the apes'; each of the hominid species thereafter will have been evolving within somewhat slower cycles, in line with the growing social and personal complexity of their lives. This will have been very closely related to the nourishment and care demands of the larger brain.

The evidence of Turkana boy, and the difficulty about establishing his age at death, may indicate the process of change. The adolescent growth spurt associated with modern humans may not have begun to occur, so it is possible that he was much closer to physical adulthood at the age of eleven than would be suggested by a straight comparison with a modern human boy with similar teeth.

On the other hand, Turkana boy's growth pattern was markedly more human-like than that of juvenile examples of previous species, and even than the preceding *Homo* species, *H. habilis*. Evidence is sparse, but Holly Smith says that such juvenile fossil material as there is suggests that *H. habilis* was markedly more ape-like in its development pattern than *H. erectus*. "There are now several individuals of juvenile *H. erectus* and they seem to show a real shift toward the human pattern of development in the teeth, and so it seems likely that the life cycle was shifting towards a human rate of growth and development." (Interview, August 1993.)

Primates take far longer to reach maturity than other mammals, because their survival strategy depends so much on learning about their social and physical environment. This is even more apparent in humans (lower strip) than in chimpanzees (top strip). Human children have to learn physical coordination and social skills and also language – which itself increases the potential for learning, as it can be used to transmit knowledge accumulated by previous generations. The long juvenile period in humans demands a huge parental investment, which itself creates pressure for long-term bonding between parents.

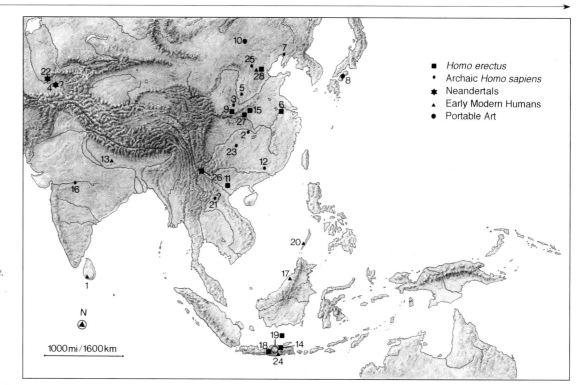

Selected hominid sites in central and southeast Asia.

1 *Batadomba Lena.*
2 *Changyang.* 3 *Dali.*
4 *Darra-i-kur.* 5 *Dingcun.*
6 *Hexian.* 7 *Jingniushan.*
8 *Kamikuroiwa.* 9 *Lantian.*
10 *Longgu.* 11 *Luc Yen.*
12 *Maba.* 13 *Mahadada.*
14 *Modjokerto.* 15 *Nanzhao.*
16 *Narmada.* 17 *Niah Cave.*
18 *Sangiran.* 19 *Solo River.*
20 *Tabon Cave.* 21 *Tam Hang.* 22 *Teshik-Tash.*
23 *Tongzi.* 24 *Wadjak.*
25 *Xujiayao.* 26 *Yuanmou.*
27 *Yunxian.*
28 *Zhoukoudian.*

Homo erectus remains have been found at widespread sites in Africa, from Swartkrans in the south to Morocco and Algeria in the north. More significantly still, they have been found outside Africa, in Indonesia and China. *H. erectus* fossils found in Java have now been dated to as early as one and three quarter million years ago, and by about one million years ago, the *H. erectus* population had spread significantly in Asia, the first hominids to do so. For the first time we can talk about hominids as something other than a local African lineage.

Robert Foley offers an explanation for this spectacular evolutionary change. "I'm not sure that any *Homo erectus* population or family or even a single brave individual ever set off to colonize the world. In fact, I'm 90 percent certain that never happened.

"But when we look at successful populations, be they human or animal, what happens is that population growth puts pressure on resources – and then populations are faced with the choice of either intensifying the way they are using the area they are currently living in, or else finding somewhere else to live."
(Interview, October 1993.)

In this scenario, the spread of population is gradual. Family groups or small communities become too large for the resources of the area they are living in. The group divides, and a new settlement or habitat is established nearby. This process happens repeatedly over time, and the geographical range of the population increases steadily, even rapidly.

This introduces the idea of distance as well as time in evolution. New things appear over time, but they also happen geographically. And in the case of *Homo erectus*, some behavior must have been present to make this development possible which did not apply to previous species or populations.

"My own guess," says Robert Foley, "is that *Homo erectus* or early *Homo* began to use more meat, to hunt more, to have more animal resources." Herbivores tend to be restricted to the plant resources in their local habitat, whereas the opposite is the case for carnivores. They can go to wherever there are meat resources of any kind – and that is practically anywhere in the globe. "So one hypothesis would be that once *Homo erectus* was a successful meat-eater or meat-scavenger or hunter, then the habitat barriers that it had previously faced disappeared."

This is a more practical, or functional, approach to the question of the move of populations than to relate it to the species' intelligence. It is not so much that *Homo erectus* had crossed a cultural threshold as that a physical barrier had been removed. It had become possible to live wherever meat could be found, so as the population grew, groups of *Homo* took advantage of the new possibilities. It now seems that from almost the first sign of *Homo erectus* in the fossil record, the species was spreading.

There is new evidence of a further practical capability *Homo erectus* may have had which would have given them an even greater freedom to move: the use and control of fire. At his dig site at Swartkrans in South Africa, Bob Brain was finding remains of *H. erectus* and of *Australopithecus robustus*, two species of hominid whose existence seems to have overlapped. At a level in the cave dated at over one million years ago, he found evidence of bones of animals which seemed to have been deliberately burnt.

There is a very clear difference between the effect of a brush-fire which may have caught an animal accidentally in its path and of a fire which has been used deliberately to cook a piece of meat. In the first case, the bone is scarcely affected by the heat, because in a sense it is protected from the fire by its surrounding meat. In the latter case, because the heat is much more intense, the bone blackens and its chemical composition changes.

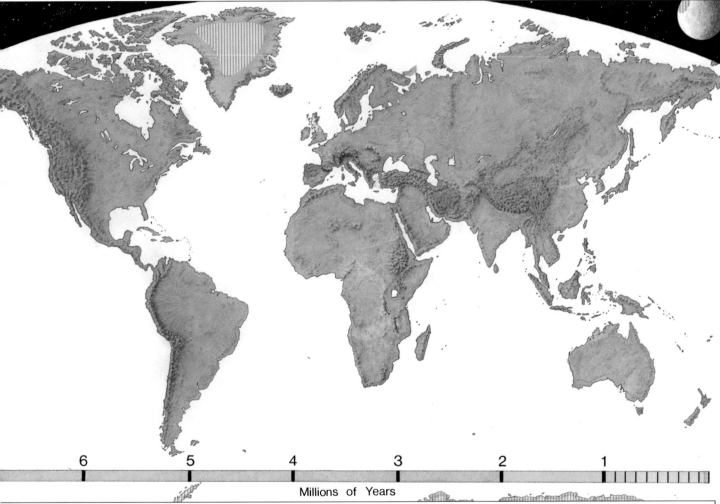

6   5   4   3   2   1

Millions of Years

...nce their divergence from the African apes about 7 million years ago, hominids were ...nfined to the savannah of sub-Saharan Africa (red area). Between 1 and 2 million years ...o a new hominid species appeared, Homo erectus, *which spread beyond Africa (orange* ...ea) as far east as Indonesia, 1.75 million years ago. By 700,000 years ago it had moved ...o warm, temperate areas around the Mediterranean. Between 45,000 and 700,000 years ...o the hominids' range increased only moderately (brown area), but from about 400,000 ...ars ago H. erectus *was being replaced by a bigger-brained and very robust hominid,* ...own as archaic H. sapiens. These reached the Thames Valley in England and other parts of Northwest Europe, but retreated southwards as an Ice Age set in. Their descendants, the Neandertals, colonized the northern latitudes permanently. By 50,000 years ago, modern-looking humans had spread far beyond their likely African homeland, and by about this time may have reached Australia by raft or boat. After 45,000 years ago there is a massive increase in range (green areas), due to the spread of fully modern humans, who replace all previous populations. Humans reached the Americas at least 15,000 years ago, and possibly even 35,000 years ago. Within the last 10,000 years, remote islands were colonized – and in 1969 hominids arrived, albeit very briefly, on the moon.

Natural fire (right) may have provided early hominids with the means of starting their first controlled fire. Above is a reconstruction of the type of hearth used by Homo erectus *populations at Zhoukoudian in China, with ash layers, charcoal, scorched and cracked stones, burnt bones, and nut shells.*

Work done by Andrew Sillen at Transvaal University on the bone remains found at Swartkrans points towards deliberate cooking. Blackening of bones only occurs at temperatures of about 400°C, and Sillen was able to show that these temperatures do not occur in natural brush-fires. In addition, the blackened bones were only being found at the one-million-year level in the cave and not at earlier levels, despite the presence deeper down of hominid and other animal remains, so something different must have been happening.

No one knows when hominids were first able to make fire, which is after all a complicated process. But it is entirely possible that for a long time prior to the discovery of fire-making they were able to make use of it by bringing burning embers back to their home base from, say, the site of a lightning strike on the savannah.

Before the Swartkrans discovery in the late 1980s, the earliest use of fire was thought to be about 400,000 years ago, a date taken from dig sites in China where the remnants of hearths had been found. The Swartkrans bones put the use of fire at almost one and a quarter million years ago – three times as long ago as had previously been believed.

Fire has another great significance apart from the ability to cook meat, which would have made the food more digestible and killed harmful bacteria. It changed the relationships between the hominids and other species of animal, especially predators.

In the Swartkrans cave, Bob Brain found that below the fire level, a high proportion of the accumulated bones, around 20 percent, were of hominids. Beyond the fire level, the proportion drops to about 5 percent. "Where the evidence for fire management is found we find a drop in the level of predation that was going on against these early people." (Interview, August 1993.)

We know that the early hominids were preyed upon by the big cats of the African savannah. They were a vulnerable species, eking out a living in competition with far more dangerous creatures. Even today, with all our lethal technology, humans have to take great care on the open African plains. Two million years ago, with no physical protection, there was a daily risk of being attacked by a predator.

The one thing the predators are afraid of is fire. The ability to set a fire would have provided a measure of protection for the first time, and would have markedly changed the balance from a subservient position towards at least survival if not dominance.

If it is true that the control of fire was available from almost one and a quarter million years ago, then the hominids had the ability to cook, to stay warm and to protect themselves to some degree from danger. This new technology would have made a very significant difference to their ability to inhabit new areas and to thrive when they found themselves in a new habitat.

Bob Brain strengthens this point when he talks about who it was who was using the fire: was it *Homo erectus*, or the other species of hominid found at the same site, *Australopithecus robustus*?

"My gut feeling is that fires would have been tended by the more intelligent of the two hominids. I would think that they were made by *Homo erectus* people, and that this process was observed by *Australopithecus robustus*. Strangely enough, the first glimpse that we have of fire making here in the cave is also the last glimpse that we have of *A. robustus*. He's not known after that.

"Could it be that the technological advantage that came to early people with the management of fire was used against these large ape-men and contributed to some extent to their extinction?

"That's the tantalizing part of this sort of palaeontology: the bits of information that you would so often like to have are simply not there." (Interview, August 1993.)

The indentations in this fossil hominid skull match exactly the position and spacing of the incisor teeth of a prehistoric leopard. Bob Brain (below) did the research at Swartkrans in South Africa which showed how much the lives of early hominids were changed by the use of fire.

The two types of australopithecine – the gracile (right) such as "Mrs Ples" or the Taung baby, and the robust (below).

Bob Brain seems to be carrying his natural diffidence and proper scientific caution a little too far. *Australopithecus robustus*, predominantly a herbivore, did become extinct; *Homo erectus*, probably a fire-using carnivore, went on to populate significant areas of the Old World. Bob Brain's discovery that the management of fire seems to have coincided with the time of *H. erectus's* population spread seems much more than tantalizing. It looks like one of the keys which enabled a major evolutionary change.

There is clear fossil evidence that the *Homo erectus* spread did take place. It is very far from clear what happened after that. Is there a direct line of descent between the diverse *H. erectus* populations and today's *H. sapiens*? Have today's Asians evolved from the early *H. erectus* population of Asia? Or is *H. sapiens* a new species which came from somewhere else, much more recently, replacing the *H. erectus* people?

This is in many ways the key intellectual debate in today's evolutionary science. And it is more than that, because it relates to the issue of the age of today's races of humans. Are modern Chinese people, for example, the product of a million-year-old lineage, or of something much more recent? Emotionally, it may be one thing to trace a common ape ancestry back to seven million years ago, and another to trace a hominid ancestry back to the African *Homo erectus* population more than one and a half million years ago. But if the most recent common ancestor of all living humans turned out to have been in Africa as recently as maybe 100,000 years ago, that seems a different concept altogether.

Those are the two major scientific positions. On the one hand, that all modern people are descended regionally from the *Homo erectus* populations, with genetic admixtures from other populations which have moved through the world at various times, responding to, among many other pressures, the moves back and forth of the ice cap. Advocates of this view lean heavily on archaeology and on the physical appearance, the morphology, of the fossil finds in different parts of the world, though not to the exclusion of genetic information.

On the other hand sits the view that in reality the *Homo erectus* population, and its regional variations, were completely replaced by a new species, *H. sapiens*, which came into being in Africa between 200,000 and 100,000 years ago. In this model there was no genetic interchange. The earlier species was, quite simply, displaced by the new, which went on to populate not only the Old World but also the entire surface of the habitable globe. Advocates of this model depend substantially on genetic research, although by no means to the exclusion of the fossil record.

There are shades of opinion in between the extremes.

Milford Wolpoff at the University of Michigan, and Alan Thorne at the Australian National University, have developed the idea of multi-regional evolution. They say that one of the first things that happens when people occupy new parts of the world is that they form geographic races. Local physical characteristics come into being, affected over time by the local climate and environment, and are passed down through generations genetically, so that a distinctive appearance begins to relate to a particular area. Physical form has evolved in different parts of the world to deal in particular with extremes of heat and cold.

People who live in very hot climates tend to be tall and have dark skin. Their large surface area helps to disperse and dispose of heat; their skin color keeps out too many harmful elements of sunlight. In colder places, people tend to be more squat, in order to retain heat, and their skin is paler in order to let into their bodies more of the beneficial effects of the lesser quantities of sunlight.

But these are not universal rules. The human species is so flexible that in different parts of the world different solutions have been reached for similar problems. Eskimos deal with cold through their size and through extra layers of heat-retaining fat on their faces. Australian Aborigines, who also have to withstand very cold temperatures for some of the year, have the ability to let their bodies drop to very low temperatures without triggering the more usual reflex of shivering.

These differences in evolutionary tactics, according to Wolpoff, are part of the pattern of regional evolution. He thinks the differences, for example, between modern Chinese and Indonesian people, and between North and South Asian peoples, reflect the differences between Chinese and Indonesian *Homo erectus* fossil specimens. A Chinese skull from 400,000 years ago has facial features – flat cheek-bones and a low nasal angle, the degree to which the nose projects from the face – similar to the facial features of modern Eskimos.

An Indonesian skull from substantially earlier, around 750,000 years ago, shows features – a sloping forehead, a more projecting face and a higher nasal angle – more like the facial shape of modern Australians. So, according to the multi-regional view, there is regional continuity from archaic to modern forms of humans.

But there are some evolutionary developments which cut across regional lines, which affect the entire species, wherever they have lived, the most obvious being the growing size of the brain.

The modern Eskimo and the modern Australian may have continued to develop separately, retaining their historic physical features; but they have both developed the larger brain characteristic of modern humans. Wolpoff thinks this can be accounted for by the fact that there has been a combination of genetic interchange between moving populations on the one hand and local stability of relatively settled populations on the other.

Genetic interchange would happen as populations encountered one another, enabling advantageous features to be spread from one place to another, helping the whole species to evolve as a complete entity. Exchange of ideas, which is unique to humans because of language, would also bring about common behavioral adaptations. But the flow of these two types of exchange would be against the background of regional continuity, in which many physical features would have been established very early in a population's habitation of a particular part of the world.

There is very wide variation in the appearance of modern humans, brought about probably by local conditions. Right is an elderly Inuk, or Eskimo, in the kitchen of his home in northern Canada; below is an Aborigine child in Australia.

If there were further moves out of Africa after the first *Homo erectus* move, the effect of these moves would have been to increase the likelihood of encounters and of genetic interchange. This is most obvious from the archaeological record in what is now the Middle East, the land bridge between Africa and the Old World, where many finds have been made which seem to indicate the movement and interchange of different populations over many hundreds of thousands of years.

The outcome of this flowing process, to Wolpoff, is modern *Homo sapiens*, which is "a way of mind, a state of thinking, a bunch of behaviors, it's not an anatomy. Evolution is not just the question of understanding the anatomy of fossils. The other key element is the behavior of the populations."
(Interview, August 1993.)

He says this in part because of the huge variety of physical type encompassed by modern humans, all the product of regional adaptation, but made consistent as a species by the big brain.

In the Wolpoff view, the most recent common ancestor of modern humans would be from among the *Homo erectus* population which spread out from Africa over one million years ago. Most of today's racial variations would have been initiated from about that long ago in the various regions of the world where *H. erectus* settled.

Milford Wolpoff has had a number of colorful public clashes with Chris Stringer of the Natural History Museum in London. Stringer rejects the idea that the ancestors of modern Chinese are the Chinese *Homo erectus*, or that the ancestors of modern Australians are the Javanese *H. erectus* specimens. He holds to the view that the ancestors of all modern popula-

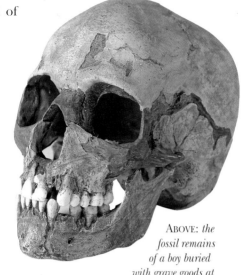

ABOVE: *the fossil remains of a boy buried with grave goods at Qafzeh in Israel are among the earliest modern humans found in Asia, dating to about 100,000 years ago.*
BELOW: *a view of Amud Cave, also in Israel, a major crossroads in the spread of early humans.*

tions everywhere are in fact a new type of *Homo*, early *H. sapiens*, who emerged in Africa somewhere around 200,000 to 100,000 years ago. This new species was so flexibly adapted that it completely replaced the existing *H. erectus* populations both in Africa and elsewhere in the world.

"We know from genetics and anatomy that modern people all over the world are actually, fundamentally, very similar. The racial differences we see are pretty superficial. I think underneath, ultimately, we're all Africans. All the racial features we find today, in my view, evolved within the last 100,000 years – probably most of them within the last 50,000 years, after the proto-modern populations got to the regions in which we find them today."
(Interview, September 1993.)

Robert Foley puts the most recent common ancestor of all living humans a little further back, though in the same basic framework, at between 300,000 and 200,000 years ago. "Which means that we all became *Homo sapiens* before we became Chinese or African or European, and therefore that what we share as *H. sapiens* is far, far more important than the minor differences that have developed since.

"There's no sense in which races, as commonly understood, are of great biological significance. The differences between human populations from one area to another are not major evolutionary differences, they are really just the minor details that have arisen in the last few thousand years." Foley's guess would be that most of the human variations visible today have arisen only in the last 10,000 years or so.

"Humans are extraordinarily plastic. Over several generations, populations can adapt to new conditions, to new environments. In the human species there is an enormous reservoir of variation which can be used as selection requires it to solve particular problems. These might be problems to do with disease, they might be problems to do with the immediate environment, sunlight, temperature and so on."
(Interview, October 1993.)

A strong line of evidence for this idea of a relatively recent common ancestor of humans has come from genetics. Inside every cell in our bodies is a nucleus, and the nucleus contains most of the DNA which scientists normally study. Outside the nucleus, however, lies mitochondrial DNA. These are small circular pieces of DNA, only about 16,000 base pairs long (compared with two billion in nuclear DNA.) They serve not to build our bodies during growth, as is the case with nuclear DNA, but act as the source of energy for the metabolism of the cell.

Mitochondrial DNA interests evolutionary scientists because it is relatively small and therefore easier to study, and it also has a very high rate of evolutionary change. So in studying closely related organisms like humans, the mitochondrial DNA can offer information about the differences between them.

Another interesting property of mitochondrial DNA is that it is passed down only through the female line. This gives a less diluted view of inherited history than nuclear DNA, in which the father and mother both contribute to the next generation's DNA structure. We know, for example, that our mitochondrial DNA must have come intact from our mother; and from her mother, and from her mother before that.

Studies of the mitochondrial DNA of living populations in the 1970s showed that there was a greater diversity among living Africans than among non-Africans. This suggests that the human lineage is older in Africa than it is elsewhere in the world. The mitochondrial DNA has been evolving and diverging for longer among Africans; so the population of the rest of the world is younger than the population of Africa.

In addition, the research showed that over all, the divergence in the world population was relatively slight – a difference of about 0.5 percent. Calibrating that difference against the rate at which mitochondrial DNA evolves brings into focus a picture of a common mitochondrial ancestor for all humans, who lived around 200,000 years ago. If that is correct, then by

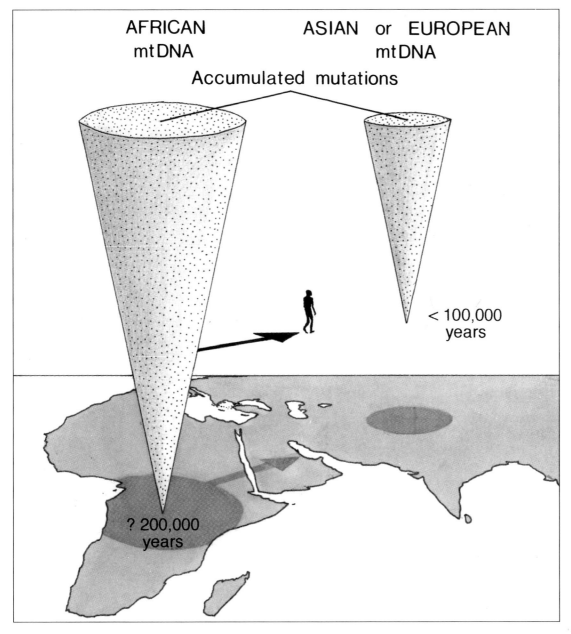

AFRICAN
mtDNA

ASIAN or EUROPEAN
mtDNA

Accumulated mutations

< 100,000
years

? 200,000
years

Mitochondrial DNA evolves quickly and is therefore useful for tracing evolution over relatively short periods of time. Researchers have found that it is far more variable in native Africans than in other groups. Assuming that the variation in mitochondrial DNA accumulates at a constant rate, Africans must have been evolving for significantly longer than other groups. This in turn suggests that non-African humans were established relatively recently, as a result of a migration "out of Africa."

definition the descendants of that ancestor must have completely replaced any previous population that had begun to diverge more than 200,000 years ago.

There are active debates about the accuracy of the molecular clock, which depends basically on measuring degrees of divergence against known dates, taken from the fossils. The molecular clock of nuclear DNA has produced a date for the much earlier common ape ancestor, about which there is relatively little controversy. But the molecular clock's ability to tie in with mitochondrial DNA and produce an accurate date for the much more recent ancestor of modern humans is less generally accepted.

Much of the original work in this area was done by the late Allan Wilson and his colleagues at the University of Berkeley in California. The notion of a common African ancestor would have caused much less of a stir if the molecular clock had dated it at one or two million years ago, because then it would have fitted in much more neatly with what was already known about *Homo erectus*. Dating it at 200,000 years ago was much more of a problem in relation to the archaeology.

The disagreement between the multi-regional advocates and the supporters of "Out of Africa," as the other view is often known, is sometimes presented as a straight conflict between the geneticists on one hand and the archaeologists and palaeontologists on the other: genes versus fossils. In fact scientists from all disciplines variously support both models, although for obvious reasons there are not many geneticists who stand behind multi-regionalism. But there are those who have difficulty with the archaeological – as distinct from the fossil – record in relation to "Out of Africa."

Lewis Binford, for example, says: "I have to say right off that from day one it was my opinion that the archaeological record did not support this argument." Binford thinks the first *Homo erectus* radiation out of Africa over one million years ago reached China, with some filtering into central Europe. He argues that there was then a second radiation which resulted in the earliest known material in Europe, dated to around 400,000 years ago, and in India, which dates to about 350,000 years ago. But he goes on: "Now, if we then come up and say: do we have any evidence in the archaeological record of a radiation out of Africa in the time period that the mitochondrial people suggest?, the answer is no, not anywhere." (Interview, August 1993.)

Chris Stringer and others would not agree, calling in the fossils to support their view. Remains of early *Homo sapiens* from Africa and Israel, Stringer would say, supply ample indications of the emergence of a new species, quite unlike the *H. erectus* people and much more like modern humans. Certainly it is clear, from a wealth of archaeological sites in Europe and elsewhere, that by about 40,000 years ago a whole new world had been born. People very much like ourselves had established settled living sites, and it is from around that time that we begin to see evidence of the great explosion of creativity symbolized by the proliferation of extraordinary cave paintings and sculpture.

At Klasies River Mouth, right on the furthermost tip of South Africa, a site is being excavated and researched which shows clear evidence of occupation, around 100,000 years ago, by people very much like modern humans. They had hearths, possibly shelters; they made tools from materials such as quartz which suggest they may have had an exchange value as well as utilitarian purposes. The scientist in charge of the site, Hilary Deacon, believes that they were able to communicate with one another much as we are able to do today.

There is little doubt that something happened which created an evolutionary convulsion, and that the answer to the puzzle lies somewhere in the evidence provided by genetic research, archaeology and morphology.

Maryellen Ruvello at Harvard University, who specializes in mitochondrial research, acknowledges the distress caused to some palaeontologists by the genetic information, but is not to be shifted from its implications. There is only enough divergence in mitochondrial material in living humans for a 200,000-year time-span back to the common ancestor. "If an older *Homo erectus* individual were to mate with an emerging new human out of Africa, then the offspring's mitochondrial type, through the female line, should show some signs of the more ancient material. And the point is that we don't see any trace of the older *H. erectus* type today."

The original mitochondrial DNA research was criticized in the scientific community on several grounds. For example, the Africans tested in the sample were American Africans, rather than African Africans. Maryellen Ruvello has dug the ground over again, with new samples and different techniques; she still arrives at the same broad conclusions, although she would put the date of the ancestor at between 300,000 and 200,000 years ago rather than more recently, and she believes that further work will need to be done in order to be absolutely certain that the originator of the species lived in Africa. But even the new research continues stubbornly to exclude the possibility of the date being as long ago as one million years, and it continues to insist that there is no genetic residue from the previous populations: the new people completely replaced, rather than merged with, the old.

There is a curious resonance in the idea of being able to trace our ancestry back to a single woman living in Africa 200,000 years ago. It is no accident that she became known as

Klasies River Mouth in South
Africa, where an early human
population made a living by
the sea 100,000 years ago.

Ape Man

A woodcut from
the Luther Bible,
showing the
traditional image
of God's creation
of the earth.

"mitochondrial Eve," or "African Eve," and that the notion captured the imagination of the general media in ways which, frankly, other aspects of palaeoanthropology rarely achieve. The fundamental conceptual clash between evolution and the Biblical story of Creation seems to have found a sort of common ground; the emergence of a whole new species from a single woman.

Of course in a sense any new species must eventually be traceable back to a single individual carrying the genetic mutation which thrived at the expense of previous patterns. And the individual was not alone in any physical sense: she was a member of a previous species as well as carrying the blueprint for a new one. So she did not come from nowhere; she emerged from an existing population as a result of the evolutionary process, the random spinning of the genetic wheel which, in the right circumstances, will produce a new species better adapted to the environment than anything which went before.

Robert Foley thinks of evolution as a process taking place through both time and space. He believes that there has been a succession of "colonization events" during the past two million years, with *Homo erectus* the first and *H. sapiens* the most recent. "Probably there were other populations, other species and sub-species which repeatedly moved across Asia and Europe out of Africa and perhaps back into Africa. In that context the 'Out of Africa' theory is just the last of many events throughout our evolutionary history." (Interview, October 1993.)

In spite of the disagreements between scientists, symbolized by the clash between the extreme versions of the multi-regional model and the "Out of Africa" model, the consistent picture which emerges is one in which there is a remarkable variety in the human species, and a degree of plasticity about our features, which are the key to our ability to populate such different areas of the globe. They do not illustrate fundamental human differences: on the contrary, they illustrate the underlying similarity of all humans.

6 _Making Images_

The idea that animals change through time and space, adapting to new conditions and environments, is not hard to grasp or to accept. Applying the idea of evolutionary selection to the human lineage provides some answers to questions about walking upright, using tools, developing large brains, and why the human population spread so far across the world.

Each of these changes has some practical need at its root. Evolution has solved, through adaptation, problems of survival, enabling the species to reproduce itself. Evolution in that sense is an ordinary process, applicable to every species.

But there is something humans do which seems much harder to fit into the evolutionary template. From about 40,000 years ago, human beings have wanted to make things which seem irrelevant to any practical requirement – none of them solves any problem that springs readily to mind – but are much prized in themselves: sculptures and other works of art, decorative items of all kinds, music.

Evolution does contain an element of chance. It would be wrong to see it as a purely deterministic process, where every new development follows a neat and logical path. In the first place there is the accidental genetic mutation, in which some new physical feature comes into being which may or may not work, and which is just as likely to be abandoned as to be passed on as a useful attribute to the next generation.

In the second place there is the fact that evolution seems to work by going one step at a time. It is not like a game of chess, where a long-term strategy is in place, based on the long-term consequences of a particular move. A change may take place which may solve one problem but will set up new difficulties for the future.

Humans are rightly proud of their creativity. It is a vital element in technological innovation – humans would not have been able to come up with the idea of the wheel without first being able to imagine what it would do. The ability to make those connections is a function of the way the human brain operates. But humans also use their creativity for a much more abstract purpose, for all those activities which can be grouped together under the general heading of "art."

Artistic expression is another of those defining human characteristics, and art provides things which we can and want to leave behind for future generations to study or enjoy. In

A cave painting from Africa, dating from about 25,000 years ago, shows three figures, apparently lying beneath a heavy bar. The paintings were found near Kandoa in central Tanzania by the Leakey family.

_135_

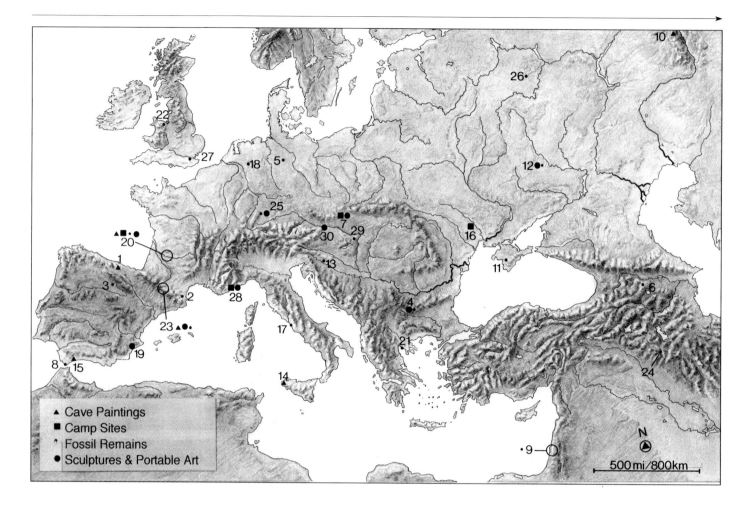

Cave paintings

1 *Altamira*, 10 *Kapovaya Cave*, 14 *Levanzo*, 15 *La Pileta*, 20 *Périgord (Font-de-Gaume, Lascaux, La Mouthe, Les Combarelles)*, 23 *Pyrenees (Niaux, Trois-Frères)*.

Camp sites

7 *Dolní Věstonice*, 16 *Molodova*, 20 *Périgord (Combe Grenal)*, 28 *Terra Amata and Lazaret Cave*.

Fossil remains

Possible Homo erectus – 6 *Dmanisi*.

Archaic H. sapiens – 2 *Arago*, 3 *Atapuerca*, 5 *Bilzingsleben*, 21 *Petralona*, 23 *Pyrenees (Montmaurin)*, 25 *Steinham*, 27 *Swanscombe*, 29 *Vértesszöllös*.

(CONTINUED OPPOSITE)

fact it would be something of a disappointment to discover that art was an evolutionary by-product, an accident brought about by other, more practically useful changes.

Robert Foley, however, points to the fact that evolution is subtle. "It is not just about survivorship in the basic sense of the environment, the temperature, the climate – it's also about surviving in a complex social world. Many of the problems we see facing humans are not necessarily the first things we would think of when we think of evolution. They're much more subtle than that, but they are none the less part of the evolutionary process, part of the environment into which humans have adapted."
(Interview, October 1993.)

Although it may be too crude to look for functions of art as though it were just another stone tool, it would be equally wrong to ignore the possibility that in one way or another it fits into the evolutionary scheme.

The first evidence of artistic activity dates back to between 45,000 and 40,000 years ago in Eastern Europe, the area we now call Bulgaria. Animal teeth have been found which had been pierced with a hole, presumably intended to be worn in the form of a necklace or sewn on to clothing.

In fact there are very widespread manifestations of prehistoric art in Europe. This does not mean that this is a purely European story; but events in Europe in the period between about 40,000 and 10,000 years ago can be taken as an example of what was happening elsewhere in the world. The detail in other places was no doubt different, but the principles of the process will have been very similar.

From those beginnings with animal teeth there followed a massive surge of creativity. Decorative work on tools and weapons, beads, three-dimensional sculptures, and of course cave painting: about two-thirds of the history of human art dates to the time of the last Ice Age.

Most people are familiar with the wonderful images from the walls of caves such as Lascaux in south-west France, which are dated at about 17,000 years ago; but the European record covers sites in other parts of France, Spain, Portugal, Germany, Italy, and Eastern Europe into Russia. The cave paintings touch a nerve in everyone. They are the only aspect of the palaeoanthropological record to have become tourist attractions. They are an unmediated contact with the distant past, requiring no explanation or interpretation on one level: the visitor can look directly at a picture created by someone probably standing in exactly the same spot tens of thousands of years ago. They work as images in themselves, and it is hard to resist the temptation to think that they were left there precisely so that we can enjoy them today.

That temptation is a classic example of laying today's assumptions on to the past. Just because today we make paintings so that they can be hung in galleries for others to admire (or buy) does not mean that the same motivation applied in prehistory.

Europe's last Ice Age began 75,000 years ago and ended 10,000 years ago. Northern Europe, as far south as London, was covered by a sheet of ice half a mile deep. South of the ice, the climate fluctuated, and the vegetation repeatedly changed between open plains and woodlands. In his book *Human Evolution, An Illustrated Introduction* (third edition, Blackwell, 1993), Roger Lewin says that these climatic variations were taking place sometimes over thousands of years and sometimes over only a few generations. The Lascaux paintings were made at about the time when the ice was at its furthest southern limit, and the climate was at its harshest.

This changing landscape, according to Lewin, was occupied by herds of horses, bison and aurochs (the precursors of today's cattle.) There were reindeer, ibex, woolly mammoths and rhinoceroses (the latter two largely in the north and east).

About 35,000 years ago, there were also two populations of people: *Homo sapiens* and the Neandertals. The Neandertal occupation of Europe can be dated back to about 150,000 years ago, and the modern *H. sapiens* have left traces from about 45,000 years ago. As from 35,000 years ago, there is no further evidence of the Neandertals.

All the artistic activity from the 10,000 years during which the two populations seem to have coexisted is associated with the *Homo sapiens* people, generally known as Cro-Magnons

LEFT: *A glacial landscape: Alsek Lake in Alaska.*

Ape Man

For many thousands of years, communities of Neandertals (foreground) must have coexisted with groups of modern Homo sapiens in parts of Europe. There is no way of knowing what happened when they encountered one another.

A Neandertal skull (left) found at La Ferrassie in France, and the remains of a Cro-Magnon man of similar age.

after the village in France where their remains were first discovered. A huge mystery surrounds the relationship between the Neandertals and the Cro-Magnons, and the disappearance of the Neandertals. This is a dramatic story in itself, and one which relates directly to the creativity of the Cro-Magnons, because many people have argued that their artistic abilities were at least symbolic of their ability to out-compete and therefore replace the Neandertals.

The coexistence of the two groups of people, which also appears to have happened in the Middle East, represents the last time in history when there were what many scientists see as two simultaneous species of hominid. The fact that one of them, the Neandertals, died out is unnerving to us today, because it is an all too recent reminder of extinction.

In the scientific world, this is a very controversial area. At its heart is a debate about whether or not the two peoples were separate species, or separate sub-species of *Homo sapiens*, or simply different races of one species. Widely varying scientific names are used for both groups which reflect the disputes.

Sometimes the Cro-Magnons are described as *Homo sapiens sapiens*, with Neandertals assigned to *H. sapiens neanderthalensis*. Sometimes the usage is *H. sapiens* as against *H. neanderthalensis*. And often, for the sake perhaps of deliberate ambiguity, they are simply Cro-Magnons and Neandertals. In order not to hang our hat on one side or the other, the latter terms are probably the safest here.

The Cro-Magnons were for all practical purposes modern humans. Although they were large and very muscular, their remains demonstrate that there is no significant difference between them and us. The Neandertals were different, and the story of their apparent character is directly relevant to the theme of creativity. Understanding who the Neandertals were, and what may have happened to them, casts much light on the subsequent success of the Cro-Magnons.

In the summer of 1856, workmen quarrying limestone caves in the Neander valley near Düsseldorf uncovered a collection of bones. They had been setting off explosive charges to get at the limestone, and their accidental discovery set off a debate on the origins of humans which continues to this day.

Darwin had not yet published his book setting out a framework for evolution. Ideas that humans may have evolved from one condition to another were set in the context of a scale ranging from the "primitive" to the "advanced." Very few early human remains had been found, and none at all outside Europe. The newly discovered bones, especially the skull, seemed to contemporary eyes to belong to a creature that was bestial and backward.

In particular the skull had a very pronounced brow ridge and low forehead suggestive, again in the context of the times, of beetling stupidity compared with the smooth and elevated brow of the sophisticated modern European. There was a widespread popular belief in the mid-nineteenth century that the character of an individual could be inferred from the shape of his or her skull, and the remains from the Neander Valley were damning on every level. There were lumps and bumps in all the wrong places, and the forehead closed the argument. Whoever the skull had belonged to was backward, ape-like, intrinsically incapable of fine thought or behavior.

Respected scientists of the time came forward with what are to our ears truly bizarre explanations of the Neandertal skull and skeletal remains. Rudolph Virchow, the founder of modern pathology, and a lifelong opponent of the very idea of human evolution, supported the view that the remains were probably those of a Mongolian Cossack soldier, a member of the Russian army that passed through Germany on its way to attack France in 1814. Features of the bones were explained because the man, as a soldier, had spent a lifetime in the saddle; on top of that, he had probably had rickets as a child, making the curvature of his bones still more pronounced. And if that were not enough, there followed a truly weird explanation of the brow ridges; these had been caused by habitual use of the muscles of the forehead – in other words, he had frowned a lot, ever since childhood.

So initially the man in the Neander cave was either some sort of ancient prehuman idiot or a Cossack.

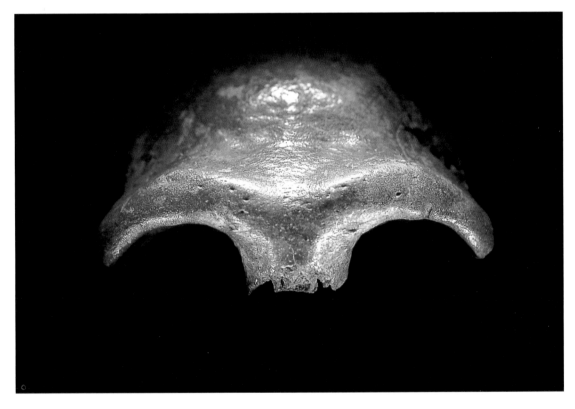

The skullcap of a Neandertal, part of the original find made in the Neander Valley in 1856.

This is one idea of how ancient sculptured objects may have been used by Cro-Magnon people in Europe 30,000 years ago. The Venus figurine is part of a ritual, perhaps to encourage fertility, conducted by senior men from the community in a special cave set aside for such purposes. Of course there is no hard evidence that events of this kind took place, and great care has to be taken not to impose today's assumptions and behavior on to the past.

Ape Man

The Cossack theory did not last very long. It might, with a stretch of the imagination, have explained some skeletal features, but it did not offer an explanation as to what the soldier had done with his clothes, weapons and other equipment, nor how he had succeeded in crawling naked into the cave which by the early nineteenth century lay sixty feet above the Dussel River.

The Neandertals have been particularly open to misinterpretation, even as the science of palaeoanthropology has progressed. It might be understandable that, at a time when there were no other fossils and when evolution in principle was not accepted, Virchow would look for some sort of explanation of the strange bones which related to recent history.

But after more fossil remains had been found, and ideas about the past had progressed, some of the later misreadings of the Neandertals seem less forgivable. Marcellin Boule, a leading French palaeontologist, analyzed one of the first complete Neandertal skeletons, found at La Chapelle-aux-Saints in the Corrèze region of France. His findings published in 1908 take no account of the evidence of osteo-arthritis in the man's spine. Boule must have been aware of the signs of the disease having been present; instead, he deduced from the skeleton that all Neandertals must have walked with a pronounced stoop.

This is how the Neandertals entered the popular imagination as the cliché "cave-man." Stupid and brutish, with long dangling arms and bent, ape-like posture, they lived their primitive, nasty lives in cold, miserable caves, spending their time bashing each other with clubs, making occasional ape-like grunts and, who knows, most probably eating each other, so low was their moral sense. No wonder they became extinct.

These uncertain views about the Neandertals persisted until very recent times. Erik Trinkaus of the University of New Mexico is probably the world's leading modern authority on the Neandertals and co-author with Pat Shipman of a recent book on the subject: *The Neandertals, Changing the Image of Mankind* (Jonathan Cape, 1993.)

OPPOSITE: *The skull and other bones of the Neandertal found at La Chapelle-aux-Saints in 1908. The thigh bones and vertebrae are deformed by arthritis.*

BELOW: *A reconstruction of Corrèze man, based on the Neandertal skull found at La Chapelle-aux-Saints in 1908. The picture shows all the signs of the early image of Neandertals as bestial and violent subhumans.*

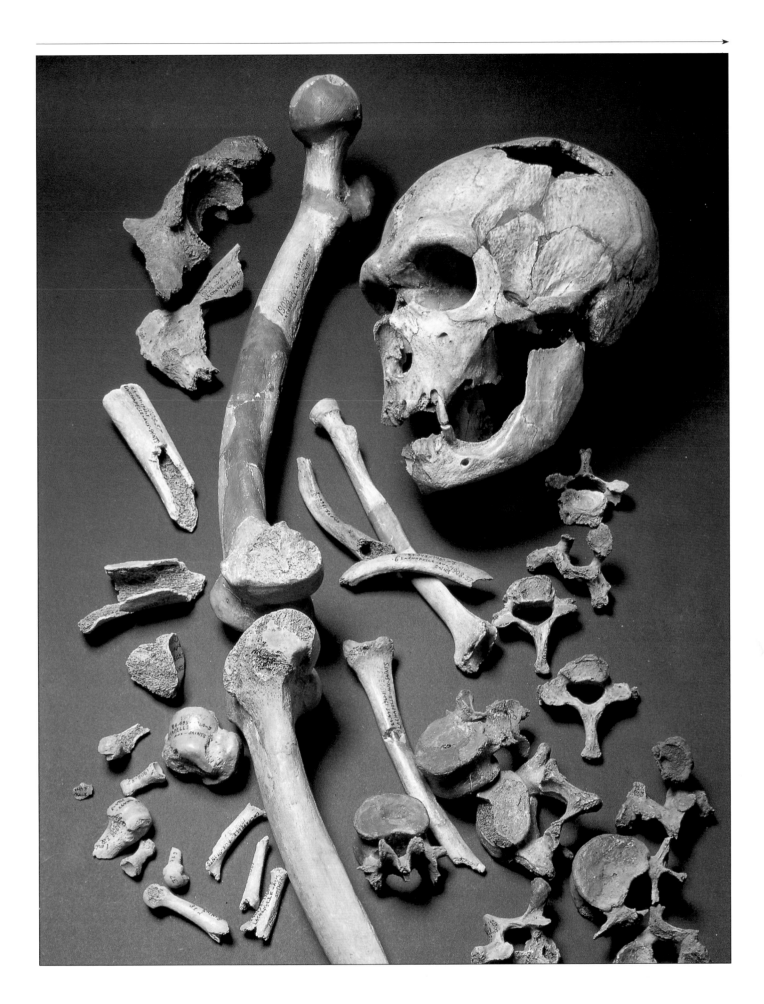

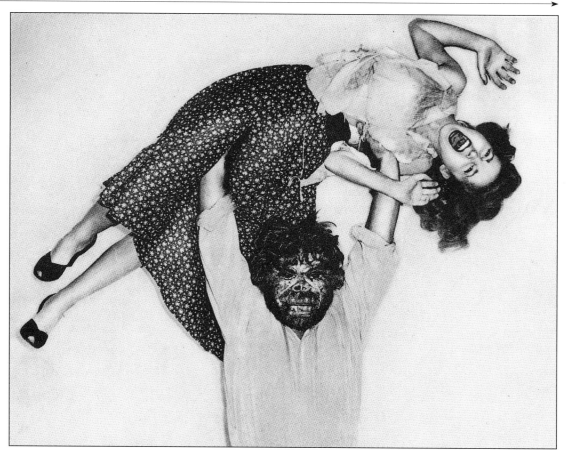

A publicity still from an adults-only cinema feature released by United Artists in the 1950s entitled The Neanderthal Man. *As usual, the Neandertals are being mercilessly libelled.*

Trinkaus identifies three phases of perception of the Neandertals. In the late nineteenth century, "they became put into a cultural or racial evolutionary scheme that was prevalent throughout the Western world at that time, where people went from periods of savagery to barbarism to civilization. The Neandertals are simply included in the category of savages: they were prehistoric savages."
(Interview, August 1993.)

Then around the turn of the century, in the period before World War I – and in the context of the Piltdown skull – the Neandertals were pushed to one side, and excluded from human ancestry. Humans during that period were thought to be so special that their ancestry must have been very long – nothing as relatively recent and yet brutish as the Neandertals could possibly be accepted as a human ancestor.

Since about the 1950s, with the discovery and identification of other much more ancient human remains from Africa and the rest of the world, "the Neandertals started to look very much like ourselves. As they became increasingly close to us in terms of ancestry, they became increasingly close to us behaviorally. But these two issues (ancestry and behavior) have to be separated. Neandertals were not either imperfect modern humans or evolved *Australopithecus* or *Homo erectus* – they were simply Neandertals.

"They were people trying to make their daily living in a very harsh landscape with a fairly primitive technology.

"The real challenge now with the Neandertals is to try to fit all this together into a reasonable picture that is both accurate for them and helps us to understand the origins of modern humans."
(Interview, August 1993.)

Just because the Neandertals may have been our ancestors does not mean that they had to be like us. Allowing the possibility that they sit in our family tree somewhere does not necessarily imply that they were nice any more than that they were nasty.

Trinkaus belongs firmly to the modern school of evolutionary science which resists fanciful extrapolation from the fossil record. What we know about the Neandertals is only what the bones and other archaeological finds actually tell us: the rest is inference.

The Neandertals were in truth fully upright; the shambling gait is nonsense. The males stood about five feet six inches tall and the females perhaps six inches shorter; both sexes were very heavily built, and it is plain from their bone structure that they were very strong.

They used tools and hunted with spears. They do not appear to have had hunting weapons like bows and arrows or spear-throwers. Instead, they used thrusting spears to fell animals at close quarters, which required great strength and, most probably, remarkable endurance.

They suffered frequent injuries. Most of the Neandertal skeletons that have been found show evidence of bone fractures which have healed during the individual's lifetime. This confirms the danger and the vigor of the life they were leading, and it also tells us that injured members of the community were being cared for in some way until they were better and could resume their normal activities.

They did indeed have the famous prominent brow ridge, and they also had a brain size similar to, or even larger than, that of modern humans. It is not clear whether they had the use of language. Certainly their brains had developed to the point where they could have had speech – but evidence as to whether they had the other anatomical adaptations (the changed structure of the larynx) is ambiguous. They had very large teeth, set in powerful jaws, which may have been used as supplementary tools.

They also had extremely large noses, up to four times the size of modern humans'. Trinkaus thinks these big noses might have been necessary to allow the Neandertals to get rid of excess body heat which would have been generated by their very heavy, muscular bodies living in a cold climate.

They controlled and used fire, and seem to have lived in the mouths of caves or under rock shelters. The average Neandertal died at the age of about thirty, but a few of them lived to a very advanced age, perhaps about forty-five or even fifty, which suggests that older members of the social groups were valued for their knowledge and memory beyond the point at which they could no longer exert their full physical strength.

The film Quest for Fire, *released by 20th Century Fox in 1981, did nothing to restore the reputation of early humans. In their popular representation, our ancestors have had a terrible press.*

There is clear evidence that the Neandertals buried their dead. The old man of La Chapelle-aux-Saints, discovered in 1908, was obviously buried deliberately. Many other burial sites, in Europe, Iraq and Israel, some of them dating back to as long ago as 100,000 years, confirm the idea that the Neandertals were the first hominids to engage in the practice.

But burial 100,000 years ago did not necessarily carry the implications of burial today. There is a big difference between disposing of an unhygienic dead body on the one hand and a ceremonial burial, with all its sense of respect and concern for the after-life, on the other.

Between 1953 and 1960, nine Neandertal skeletons were excavated from the Shanidar cave in Iraqi Kurdistan, of which five seemed to have been intentionally buried. It transpired from soil samples from the excavations, carried out by Ralph Solecki, that a much larger concentration of wildflower pollen was present in the vicinity of one of the skeletons, known as Shanidar 4, than would have happened naturally. Solecki felt this suggested that the elderly man in question could therefore have been buried along with offerings of wild flowers; in tune with the times, he published a book in 1971 entitled *Shanidar, the First Flower People*. But the possible deliberate presence of flowers at one grave site in the Shanidar cave is not a pattern of behavior, and there is no other known example of Neandertals being buried with any suggestion of ritual or of grave goods of any kind.

The other thing which is certain about the Neandertals is that their characteristic features disappear from the fossil record as from about 35,000 years ago. There seems to have been a progressive east-to-west diminution of their population, until finally it fades out altogether.

Most scientists would agree about how the Neandertals began. They were probably the descendants of part of the *Homo erectus* population which spread out of Africa over one million years ago, and subsequently adapted locally to local conditions, especially to the constraints of life in the Ice Age. But scientists do not agree about what happened at the other end of their time-span.

To return to what is known: for as much as 10,000 years and probably more, the populations of Neandertals and Cro-Magnons overlapped in Europe. As from about 35,000 years ago, the Cro-Magnon/*Homo sapiens* population went on to thrive, and eventually to occupy the whole world, and the Neandertals vanished.

The detail of exactly what happened to the Neandertals is simply not known. There is no evidence of any kind of physical clash between the two populations. It is not as though there were a Cro-Magnon blitzkrieg through Europe. There are two main views on this.

OPPOSITE: *The partial skeleton of a male Neandertal, found at a burial site at Kebara, Israel, and dated to about 60,000 years ago.*

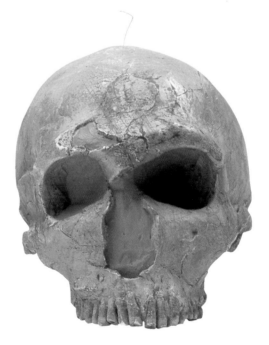

FAR LEFT: *an early modern human found at Predmost in the former Czechoslovakia.* LEFT: *a Neandertal from La Ferrassie in France.*

One is that the Neandertals were not in fact a separate species, but interbred to a greater or lesser extent with the incoming *Homo sapiens*, whose genes eventually became dominant at the eventual expense of the genes delivering Neandertal characteristics.

The other is that the Neandertals were a separate species and that their birth-rate was slower than that of *Homo sapiens*; they were out-competed and simply replaced, within relatively few generations, by the more flexible and technologically more advanced *H. sapiens*.

Which of these scenarios accurately describes what happened will probably never become absolutely clear. Both have their advocates, who broadly speaking divide along the lines described in the previous chapter: the proponents of a multi-regional evolution against those who support the idea of a new species originating in Africa within the last 200,000 to 300,000 years.

From one end of the spectrum and drawing on a much more recent illustration, Milford Wolpoff says: "The extinction of the Neandertals is cultural, not physical. New ideas came into Europe and the consequence of this was a coalescence into really new forms of human adaptation.

"What happened to the Neandertals is what happened to the Tasmanians. There are literally hundreds of thousands of people today of Tasmanian descent, but there's no more Tasmanian culture left, and that's the sense in which the Tasmanians are gone. The Neandertals went the same way. We know they were there, their descendants are there, their features are there today – but they clearly were in the process of change and they became the ancestors of modern humans."
(Interview, August 1993.)

Chris Stringer replies: "I think it was a question of competition for resources. The way of life of modern humans allowed their populations to grow, and with limited resources the Neandertal populations were under competitive pressure, and I think gradually their numbers just dwindled. I think it was just attrition. Their numbers just faded away. The fossil evidence for me shows very little evidence of a crossover of the distinctive features of Neandertals and modern people."
(Interview, September 1993.)

Erik Trinkaus thinks that there are probably Neandertal genes in today's European population. Indeed, as an American of European descent, he says that he would be proud to find Neandertal genes in himself. Perhaps he feels some sympathy for the battering their reputation has taken over the years.

But he is certainly of the opinion that the scientific argument about whether or not Neandertals and *Homo sapiens* are separate species is somewhat irrelevant, because a technical debate about the nature of species conceals a proper understanding of the process of change.

"The question is, why did the Neandertal anatomy and behavior pattern disappear, to be replaced by that of early modern humans? The question can be rephrased, not in terms of why the Neandertal pattern became extinct, but why the early modern human pattern was so much more successful.

"The answer is probably reflected best in archaeological material suggesting a much more complicated social system, and certainly all the personal ornamentation, much more elaborate burials, and organization of activities that was socially encoded enabled those early modern human populations to be more successful in the long run."
(Interview, August 1993.)

A Cro-Magnon skull and mandible found at Predmost in the former Czechoslovakia, and dated to about 30,000 years ago.

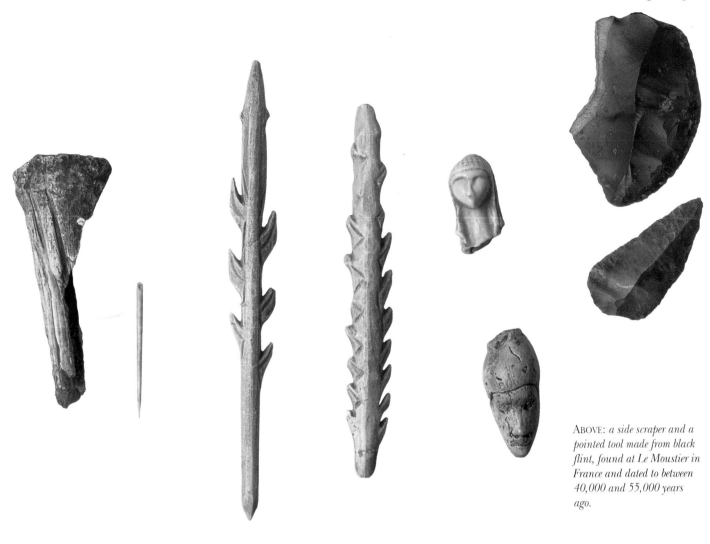

ABOVE: *a side scraper and a pointed tool made from black flint, found at Le Moustier in France and dated to between 40,000 and 55,000 years ago.*

The Cro-Magnons had better technology, better weaponry, more effective shelters. They were able to keep fires going for longer, radiating heat more efficiently. Trinkaus elaborates: "We get a series of changes in behavior that are probably reflections of an overall change in organization and the way people went about doing things that made the new pattern more successful.

"And many of the other features we see, like more lightly built bodies, as well as less wear and tear on the body, are secondary reflections of this more efficient system – and eventually that allowed them to get by with fewer calories."

According to Randall White at New York University, whose field of study focuses on art and ornamentation, the period of overlap in Europe saw the Neandertals beginning to produce some personal ornaments and beads. But there is no sign at all of them doing this before the Cro-Magnons arrived on the scene. Indeed there was very little change, through the whole Neandertal period, in their general technology.

The tools the Neandertals were using 150,000 years ago were almost exactly the same as those found from 40,000 years ago. "There's more technological change in the first 5,000 years of the Cro-Magnons in Europe than existed in all of previous human evolution on the European continent.

ABOVE LEFT: *from left to right, a bone used to make needle blanks, a bone needle, a harpoon head and barbed point carved from antler, and two heads carved in mammoth ivory. All date from between 18,000 and 30,000 years ago.*

151

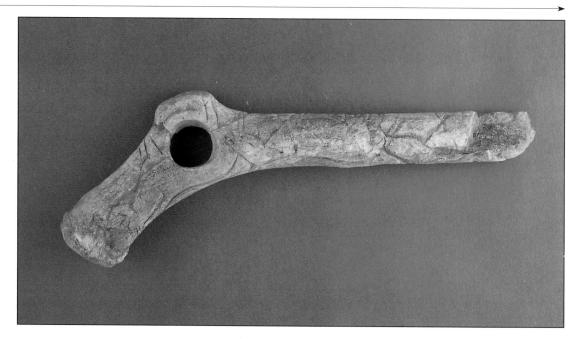

*Right and on facing page,
ornaments made from antler.*

"There's a whole new framework that's emerging for innovation and change in the technological world. We see continuous and regular changes in the form that tools made of things like bones, antler and ivory take and in the techniques that are used to work with them through the period from 35,000 years ago to about 10,000 years ago.

"There is continuous technological change, adapting to new circumstances, in some cases change almost for the sake of change, in direct contra-distinction to what we see among Neandertals."

(Interview, August 1993.)

No one can objectively describe the Neandertals as unsuccessful. They thrived in Europe for over 100,000 years, or about 4,000 generations. But from the time of the appearance of the Cro-Magnons, the pace of change accelerated remarkably.

The change for the sake of change which Randall White alludes to is probably the first occasion in human evolutionary history that this idea has surfaced. "It wouldn't surprise me that in many cases the explanation for these changes is simply that something new came up that became interesting and attractive and a challenge – and people responded to it."

This does not mean, however, that the first decorative objects were made for no reason. Connections are sometimes made between the first manifestation of artistic activity and the emergence of fully complex language; the use of language to express ideas and conjure up word pictures is related to the further step of actually making the pictures, in either two or three dimensions.

If that were the case, it would be difficult to explain the relative absence of ornamental material from the period of 100,000 years to 40,000 years ago. There were almost certainly modern humans living in southern Africa, if not elsewhere as well, by over 100,000 years ago. The excavated settlement at Klasies River Mouth has signs of all the modern activities associated with early *Homo sapiens* – except the kind of art objects which do not appear until 60,000 years or so later in Europe.

The Klasies people were making more sophisticated tools than the much older stone tools of East Africa, and were making them from somewhat different materials, some of which, including quartz, seem to have been chosen for their appearance as well as their function. But they left no beads behind, and no apparent ornamentation on their tools.

On the other hand, there is little doubt that the Klasies people had fully developed language, and this suggests that some other factor needs to come into play in seeking the driving force behind ornamentation. Randall White offers an explanation.

"I tend to think that it's not coincidental that it first happens with the expansion of these populations into new continents and into new areas. It may in fact have something to do with their encounter with already existent populations that they had to deal with. They had to communicate something of social identity towards people who already occupied these regions, that is, the Neandertals."

Anthropologists of modern societies point to the use of all kinds of personal adornment and decoration as methods of identifying the wearer as part of the social group. Be it the ornate tiaras of royalty, or military decorations for formal dinners, or studs in the ears of punks, codes are used to declare who we are.

"So in studying this kind of [ancient] material," says Randall White, "we might imagine that what was going on was the first time in human evolution that we have the internal subdivision of human societies into different categories of social persons."
(Interview, August 1993.)

At first sight it is surprising to learn that three-dimensional art came before two-dimensional representation. At least 5,000 years separate the first sculptures from the first two-dimensional pictures in the form of engraved animals. Perhaps it was easier to reduce one three-dimensional object – a horse, for example – to a smaller-scale model of itself than it was to work out how to give a sense of a three-dimensional object on a flat surface. Perhaps it required a technical advance at the same time as an advance of the imagination, in the sense of the realization that it was possible.

Randall White points to the great difficulty of understanding the emergence of particular types of object and the way in which they were treated or used. Female statuettes appear from somewhat over 30,000 years ago. In some cases they have apparently been deliberately broken in two and the pieces separately buried. "This is telling us something about the ritual context for these objects. Probably a complex set of beliefs and ideas has emerged in association with this early body of symbolic material."

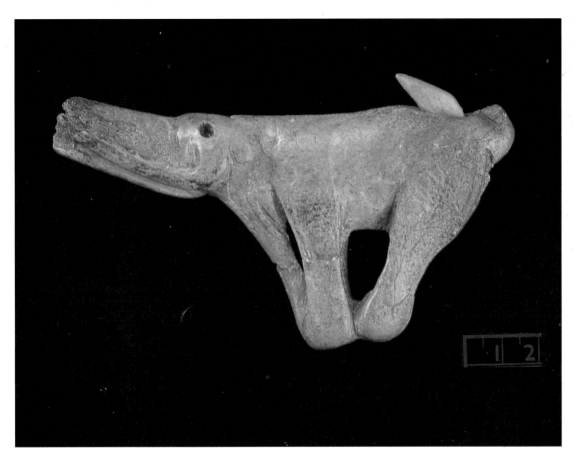

By the sea at Klasies River Mouth in South Africa, 100,000 years ago, there lived a community of modern humans whose artifacts were often made from special materials, suggesting they were used for exchange or had some other symbolic importance in addition to their practical uses.

Female figures from 22,000 to 30,000 years old. From the left, a figure of fired clay found in Moravia; one of mammoth ivory found in France; a limestone figure found in Austria; and a figure of mammoth ivory found in the Ukraine.

The Venus of Willendorf is the name given to a famous female statuette found in Austria and dating to over 30,000 years ago. Perhaps unfortunately, such statuettes have all been labelled "Venus," a practice apparently started by the Marquis de Vibraye in the 1860s in connection with his "Vénus impudique" from Laugerie Basse in France. Much effort has been expended trying to work out why most of the Venus figurines seem to have obesity, sagging breasts and prominent buttocks in common.

Paul Bahn, another world authority on prehistoric art, tells in his book *Images of the Ice Age* (Facts on File, 1988) – which is illustrated with the wonderful photography of Jean Vertut – of a study of 132 such Venuses which demonstrates that in reality they depict a range of sizes, ages and shapes of women. It is merely that the more extreme examples of shape are the ones most often seen and used for illustration. And, of course, the idea that making the Venuses was a common activity has to be seen against the fact that they were made over a time period of upwards of 25,000 years – and that it is we who have chosen to group them together under a single umbrella name.

The Venus of Willendorf has another curious characteristic which is yet to be explained. She has a kind of cap, or perhaps it is her hair, covering much of her face. What significance this may or may not have also needs to be seen in light of the undoubted fact that the makers of these objects must, over the years, have varied in their technical ability.

One clearly interesting feature of the Willendorf item is that it was made from a particular type of limestone not found anywhere in Austria, so either the object itself or the raw material must have been brought in from elsewhere. It also had traces of red ochre on its surface, which indicates that it must originally have been colored.

When prehistoric art of all kinds – both the portable sculpture and the cave paintings – was first discovered, it was widely believed that this was art for art's sake, that it was simply there to be admired. Then, by the mid-twentieth century, the view developed that art for art's sake was a modern concept which could not necessarily be applied to ancient objects, and explanations were sought which treated every object or image individually rather than grouped together in the again modern idea of "art."

The female statuettes have often been interpreted as having been used in fertility rituals. The difficulty with this idea, according to Randall White, is that in today's hunting and gathering societies rituals are often to do with reducing rather than increasing fertility. Increasing the population is easy, but keeping it at an acceptable level requires extra help.

Set against this is the fact that even in today's world, with its serious problems of overpopulation, some religious groups still see population increase as a highly desirable goal. The need to increase fertility may be more powerful than the most rational desire to maintain it at a tolerable level.

The care that was taken 35,000 years ago in the making of beads and figurines is quite remarkable. Randall White has conducted experiments in real time in an attempt to work out the processes that were used and the time that was taken.

In the earliest times the most favored material for beads was ivory from mammoth tusks. Even the initial process of breaking out a small enough piece of ivory from which to make a bead is far from easy, and then the processes of shaping it, making the hole to thread it, and polishing it, all need great care and take substantial amounts of time.

Ivory in its raw state is not a particularly attractive material, and it takes hard work to give it the smooth sheen characteristic of the finished product. Red ochre is a natural mineral, and works as a natural abrasive. Studying ancient beads through a scanning electron microscope reveals the polishing marks only at very high levels of magnification, which suggests a very fine abrasive surface, exactly as provided by red ochre, perhaps when rubbed on to the bead with a piece of animal skin.

Randall White's practical experiments show that it would have taken between one and three hours to make a single polished bead from a small piece of ivory – and the beads were made in their hundreds. There must have been very good reasons to devote so much time to making beads, in an era when the practical needs of obtaining food and maintaining shelter were obviously also very time-consuming.

Other materials used for beads were animal tooth, seashells and soapstone. All these materials, as well as ivory, are capable of being given the soft sheen characteristic of the prehistoric beads. Even beads which have been discovered underground after 30,000 years still retain their ochre-impregnated surface and traces of ochre deposits which have been left behind in the thread-hole from the polishing process.

White believes that beads were often used on clothing rather than simply being strung together as necklaces. "If you graph the dimensions of these beads, you'll find that they fall within an extremely tight range of variation.

"This probably means that they were part of what I call larger ornamental constructs, they were applied in a systematic way to compositions on clothing. One can imagine a tunic, for example, made out of animal skin, on which perhaps hundreds of these beads were applied. With the microscope we think we can see traces of them having actually been sewn on to clothing."
(Interview, August 1993.)

A human head carved in mammoth ivory from Brassempouy, France, dated to between 22,000 and 30,000 years old.

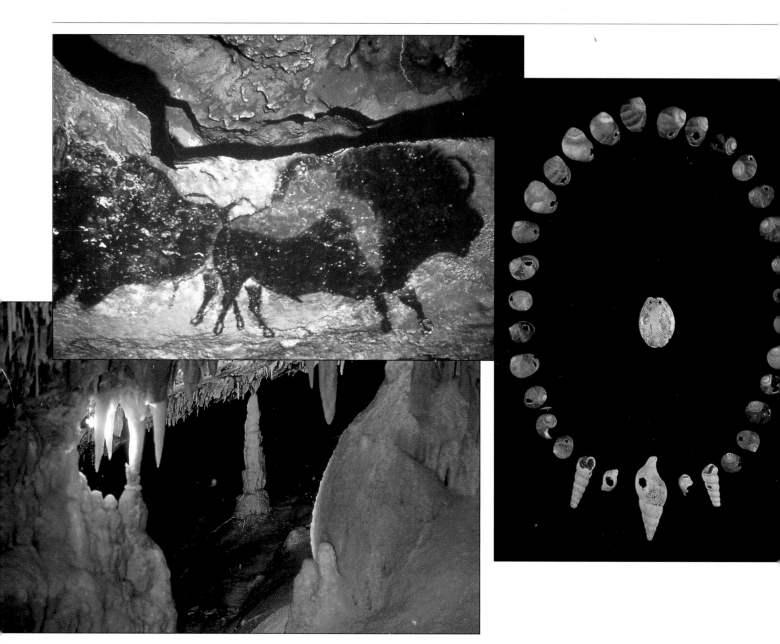

Images from the caves of France. Top, and lower right and left on the facing page, paintings from Lascaux. Lower left on this page and upper right on the facing page, the recently discovered underwater cave of Grotte Cosquer. Far right on this page, a seashell necklace, designed to be handed down through the family.

Another key to the possible use and meaning of beads and figurines, as well as to decorative work on the hafts of tools or spear-throwers, is the fact that they were durable. They are permanent items, which can be passed down through the generations, adding a long-term character to the sense of identity they may have given their immediate owners.

Terrence Deacon, the neuro biologist at Boston University, puts the beginning of durable art at the time when it became more usual to live in settled camp-sites than constantly on the move. Modern nomadic societies have forms of artistic expression such as dance, song and body decoration, but these are all ephemeral. Only when a more settled life-style comes into being do durable items start to be made, which otherwise would have been impractical to carry from place to place as the people continued their search for food and water.

Passing the items on through a family or group of families must have transmitted memories and stories that were important to those people. To us, they are the remains of a culture which we can only guess at, and most of the guesses involve superimposing our own culture and our own package of assumptions about what is attractive and meaningful on to a world which we simply do not know.

Similar principles apply to cave art. Since it was first recognized as prehistoric, a great deal of work has been done on the meaning of the images on the walls and on the tech-

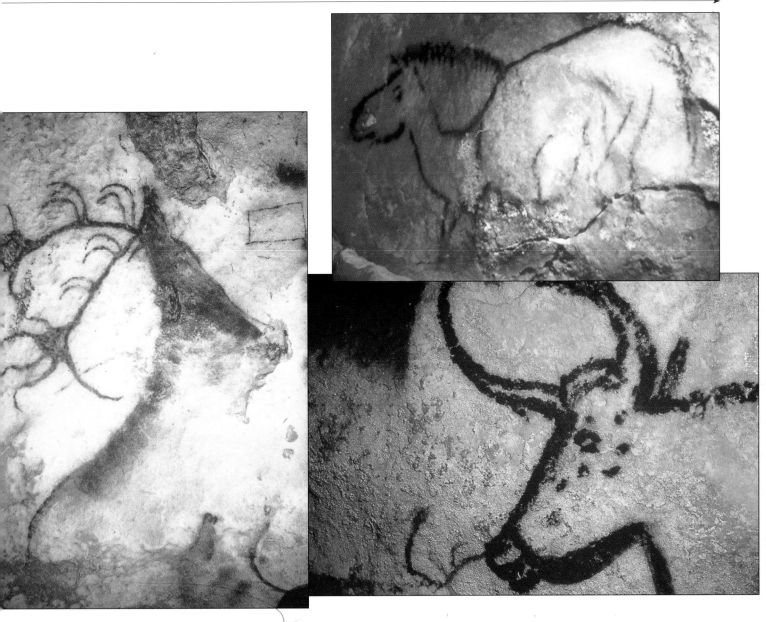

niques used to make them. The first claim for prehistoric cave art was made at Altamira in Spain by Marcellino Sanz de Sautuola in 1880. His daughter Maria had spotted pictures of bison on the cave ceiling while de Sautuola himself was looking for sculptures and other artefacts on the cave floor. But it was not until two decades later, following finds at the French sites of La Mouthe, Font de Gaume and Les Combarelles, that his claims were taken seriously.

The greatest concentration of decorated caves is in south-west France (the Périgord and the Pyrenees) and in Cantabrian Spain, although there are more scattered examples in Italy, Portugal, Italy, Yugoslavia and Romania. Lest this should be thought an exclusively European story, it is important also to know that there are decorated caves in South America and Australia, ranging in age from 24,000 to 10,000 years.

The images on the walls are usually those of animals; very few of the pictures involve humans. There are no ground-lines and there is very little by way of background or scale between the various animals. The earliest images were made by engraving shapes in clay, probably with the fingers, and the creators used the natural shapes they found in the caves to adapt into the forms they wanted. Shapes in the ceiling at Altamira are adapted into the bodies of bisons, and stalagmites and other shapes are often used as parts of figures. The clay

Bison fashioned from clay, found in the cave at Les Eyzies in France.

caves of the Pyrenees have rare three-dimensional sculptures, including a wonderful bison at Le Tuc d'Audoubert.

Most of the animals depicted are horses, bison, aurochs (wild cattle), deer and ibex. There are also mammoths, bears and other big carnivores. The most common representations of humans are the outlines of hands, frequently those of women or young people. These "negative" images of hands sometimes have fingers missing, either because of mutilation or, we assume, because the deliberate withholding of a finger had some special meaning.

It is very hard to understand why particular pictures appear in particular places. One very persistent idea has been that they related to hunting, either symbolically, to encourage good fortune, or more practically as an educational device, to record the features of animals which provide good resources in the area or even to give instruction in hunting techniques. But this argument does not really stand up to systematic analysis; the animals depicted on the walls rarely bear any relationship to the bones found in those caves, left behind by those who ate their meat. And there are virtually no pictures which include tools, weapons or chase scenes.

A huge study by two French scholars, Annette Laming-Emperaire and André Leroi-Gourhan, published between 1957 and 1965, investigated the pictures not as individual images, as they had tended to be seen before then, but as combinations of images relevant because of their layout and relationship to one another in the topography of the cave. The work is described in detail in Paul Bahn's book. Although some of the conclusions do not appear to have stood up to further testing, it is clear that there is a repeated pattern, running through all the cave sites, in the relationship between horses and bison.

The mere fact that such patterns can be detected hints strongly that the cave paintings were being done in line with a consistent set of ideas, perhaps a belief system, and that they

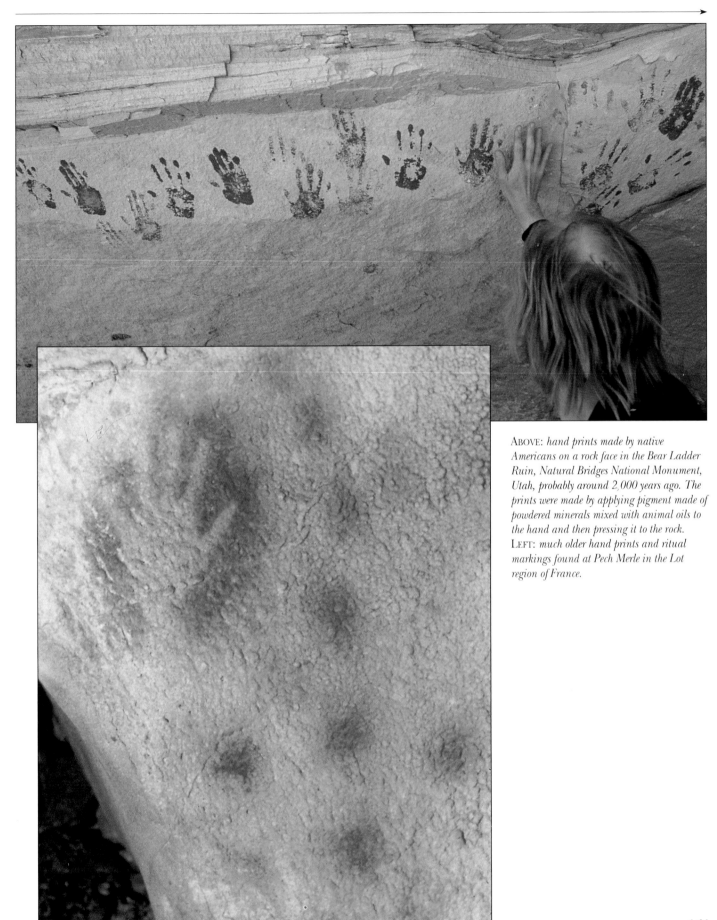

ABOVE: *hand prints made by native Americans on a rock face in the Bear Ladder Ruin, Natural Bridges National Monument, Utah, probably around 2,000 years ago. The prints were made by applying pigment made of powdered minerals mixed with animal oils to the hand and then pressing it to the rock.*
LEFT: *much older hand prints and ritual markings found at Pech Merle in the Lot region of France.*

The Venus of Galgenberg, a dancing woman carved from green serpentine, found in Austria and dating from about 30,000 years ago.

were not simply a set of images painted more or less at random over time. Randall White says: "The caves were organized. The process of painting them consumed a lot of effort, labour and planning. People weren't just painting whatever came to mind wherever it came to mind. There was a place for bison, there was a place for deer, there was a place for carnivores, and there was a place for humans in the spatial organization of these things."

In Lascaux, archaeologists discovered twelve different colours of pigment, only four of which occur in nature. Randall White relates that physicists have established that colors were achieved by heating natural pigments to temperatures of $1,000°C$. Some of the pictures could only be made by erecting scaffolding, and Lascaux again provides evidence that this must have been done to reach painting services twelve to sixteen feet above the cave floor.

Again, many of the caves were in pitch-darkness, so lamps must have been used to provide light. Lamps have been found which were made by creating a hollow depression in the middle of a small slab of stone and burning animal fat in it, probably fat from horses, bison and deer. They give off much the same light as a candle today; more than seventy-five such lamps have been recovered from Lascaux, where they appear to have been stacked against a wall.

Randall White believes that people might have used a lamp to get inside the cave, and once inside would have left it there, returning on the next occasion with more fuel. "It's entirely possible that they were lighting up significant areas of the cave. It's unlikely the scaffolding could have been built without at least a fair amount of light being available for its construction."

It seems that the painted caves were not the caves where people lived. And it seems from footprints that have been left behind that members of the community of all ages went into the caves. Randall White has seen the footprints of a two-year-old child deep in the cave of Le Tuc d'Audoubert, with its toes clenched at the edge of a two-foot drop. This was "800 metres underground in the darkness, having to have got there by crawling and scrunching his or her way through the tiniest of cavities along with the adults that accompanied it."

In some way the caves might have been places of sanctuary, or places to hold knowledge about beliefs and traditions. "What may be being imparted to young people and future generations is information about how animals are, what they are spiritually, cosmologically or mythologically. What the primary driving myths of the group are."

In the same way that we do not just look at the images on the walls of churches, but absorb the whole experience of being in a church, it must be the case that the paintings in caves were part of an overall experience. Caves are dark and frightening places, with strange noises and a very special atmosphere. Looking at reproductions of the cave paintings can only convey a small fraction of the feeling of being in the cave with the painting, and it requires a leap of the imagination to place oneself there as it might have been 18,000 years ago. But such an imaginative effort is needed to begin to get to grips with the overall experience.

If the cave paintings were to do with a belief system, then presumably no one will ever know exactly what it was, just as we will never know for sure what was being imparted to the future through the figures and beads. But we can say, and we know this from the archaeological record, that in the break between the way the Neandertals and the Cro-Magnons lived lay a whole set of new ways of thinking about the world and its meanings.

7 *Still Evolving*

Until about twenty years ago, the science of evolution was dominated by star palaeontologists and star fossils. Theories about human evolution, often based on the search for a key missing link between the apes and humans, were built by visionary individuals around individual finds.

Raymond Dart and the Taung baby; Robert Broom and Mrs Ples; Sir Arthur Keith and Piltdown Man; Eugene Dubois and Java Man; Louis Leakey and *Homo habilis*; Richard Leakey and KNM-ER 1470; Donald Johanson and Lucy . . . the names are linked together, the finder and the found. Sometimes the theories built around the great discoveries were right – Dart's *Australopithecus africanus*, for example – and sometimes, of course, they were wrong – Piltdown, most obviously. But often their evolutionary significance has been clouded by a screen of debate about whose fossil was oldest, most human-like and therefore the biggest and best.

The driving purpose of about the first 100 years of the science was to trace human ancestry in such a way as to reinforce the idea of human uniqueness. The assumption was that because humans in the present are so extraordinary compared to other animals, they must have had an equally extraordinary past; there must be a moment at which these remarkable qualities came into being.

In fact the core of Darwin's thinking, and his revolutionary contribution to human history, was to put humans on the same level as every other species, and to say that the fundamental evolutionary process was exactly the same for all species. Yet however clear and simple this thought may have been, humans have never quite wanted to accept it. As Robert Foley wrote in his book *Another Unique Species* (Longman, 1987): "Darwin may have shown that we are descended from more lowly forms . . . but in our hearts we know that humans are protected from the onslaught of the common herd of animals by an unbridgeable chasm. The walls of human uniqueness are in pretty good shape after more than a century of Darwinian battering."

The walls have been kept in place, in part, by concentrating on the importance of the brain and its development. The Piltdown hoax was able to succeed because the prevailing

ABOVE: *Robert Broom, finder of "Mrs Ples" at Sterkfontein in South Africa.*

OPPOSITE: *Samples of lunar material being collected on the Apollo 16 mission – raising the possibility that one day, humans might colonize other planets, with unforeseeable consequences for the evolution of the species.*

165

Ape Man

On this page, an upland agricultural village in rural China. On the facing page, a Bushman woman in the Kalahari in southern Africa. Across great expanses of the modern world, the norm of daily existence is not city life but a subsistence-based, day-to-day economy in the open countryside.

view was that the development of a big brain was at the root of all human development. The brain was the origin of humanity. The brain created human culture.

It is easy enough to see why the history of human development became intertwined with the history of the human brain. The product of the brain is the most obvious manifestation of human uniqueness. It is more exciting than upright walking, which seems merely anatomical; the brain is about the mind, the spirit and the soul. And this concentration on the brain has led to two extremes of opinion.

On the one hand is the view, espoused by the early twentieth-century scientists who supported Piltdown, that the brain enabled humans to evolve to their present superiority and control. They were able to take charge of their own evolution and promote themselves to a better status. It gave them a runway from which they could take off and diverge from all other animals on an intrinsically separate route. This idea lies behind the instinctive desire to set the origin of humanity as far back as possible in history.

On the other hand, accepting the notion that brain development was relatively recent, lies the sharply contrasting view that nothing which went before is of great interest or importance. In his famous book *The Ascent of Man* (BBC, 1973) Jacob Bronowski says that the biological history of man occupies millions of years, but that his entire cultural history is crowded into 12,000 years, "which contain almost the whole ascent of man as we think of him now."

Bronowski claimed that man's cultural history effectively began at the end of the last Ice Age, when a more settled existence combined with the beginnings of cultivation and agriculture to create a fundamental change: the decision to give up a nomad existence in favor of village life. "I believe that civilization rests on that decision. As for people who never made it, there are few survivors . . . Civilization can never grow up on the move."

Today's scientists no longer support the brain-first theories of the Piltdown distraction. They would be unlikely to accept Bronowski's hard break between biological and cultural evolution. They would see a much longer process of development of human culture; and they would link biological and cultural change, seeing the two as at least dependent on each other and very possibly part of the same evolutionary process. Most of them would say that there is no single origin of what we refer to as humanity, and that evolutionary science should not get caught in a sealed valley looking for it.

Every species as it evolves acquires special abilities, adaptations to its way of life. Humans are no different, although we like to think our special qualities are even more special than those of other animals.

At last, it seems that the walls Robert Foley describes as protecting the idea of human uniqueness are coming down. *Homo sapiens* looks more and more like a species of primate which has been evolving in the same way as every other species but which has acquired some remarkable properties along the way.

Rick Potts of the Smithsonian Institution, who has made a lengthy study of the very rich archaeological site at Olorgesailie in Kenya, says he is often asked when man first appeared on earth. "And my answer to that now after many years of study is that it depends. It depends what you mean by human." Upright walking began more than four million years ago, tool-making two and a half million years ago. "Other people will name different things, large brains, or art. Human uniqueness has been a long process, and that's what evolution is all about. It's a lengthy transformation from what we construe to be apeness to what we construe to be humanness."
(Interview, August 1993.)

Different aspects of our biology and our behavior evolved at different times. There is no mystical moment at which we became human. As Robert Foley says: "We accumulated during our evolution a bundle of characteristics which we now think of as being human.

"Nor is there any sense in which this bundle of characteristics was accumulated in a ladder-like progression leading inevitably towards our own species."
(Interview, October 1993.)

We know that there have been many hominid species, and as fossil excavation proceeds, there may turn out to have been more. We know that all but ours are now extinct. We also know that the bundle of human attributes was acquired unevenly, by different species, and that some species, which had some of our features, were not on our direct ancestral line.

Despite continuing scientific disagreements about important individual issues, much of the broad outline of the picture has been clarified. Most of the fundamental questions about when key changes took place, and the order in which they happened, have been answered by fossil and molecular evidence.

This knowledge puts humans in a new context. The idea of evolution should make us humble, not arrogant. As Foley says: "We've got this paradox that we're very special, but the process by which we've become special is itself an ordinary process that occurs in every species. That's what makes this such a difficult and exciting discipline to study, and also, of course, that's why it's so disconcerting.

"On the one hand we want this uniqueness, we want to say we're a new type of animal – but on the other hand we're not, we've got the same ancestry, the same biological mechanisms, the same basic structure as any other species on this planet."

Given much more certainty about what has happened and when, today's scientists are now concentrating on answering the question "Why?" And that debate is intensely relevant to the present and the future. If we understand why changes took place in the past, we have a better chance of understanding what might happen to us.

There are two distinct models of evolutionary change. In the first, it is possible to discern gradual, steady progress. Language, for example, would be improving over time, becoming more sophisticated in stages. The australopithecines would be communicating in much the same way as today's chimpanzees and gorillas; evolutionary selection would operate on the various later hominid species, leading eventually to today's complex spoken and written language. There would be a direct progressive correlation between a species' ability to use language and its chances of avoiding extinction; language would become an essential weapon in the survival stakes.

This is a tempting idea; it fits, after all, the day-to-day experience of learning. Over time, we get better at doing things, and that puts us in a stronger position to succeed at whatever we are doing. Modern humans are highly successful whereas all other hominid lineages are extinct – they must have failed in some aspect or other of their lives.

The other model of change stands rigorously back from words like progress and advance. In this model you hear few phrases about the onward march of humanity, and you see much less emphasis on tracing the genealogical tree of *Homo sapiens*. What you do see are attempts to explain the circumstances which have brought about change, which give less weight to the hominids' own efforts and capabilities and much more to the effects of the environment in which they lived.

And the key here is the knowledge that evolution is an imperfect process. It does not provide complete solutions to problems and present a total package at any one instant of time in which a species has all the attributes which it requires for a given environment.

We know about evolutionary compromise: upright walking has caused anatomical problems; the big brain creates huge demands in terms of nutrition, child care and social organization; speech requires a larynx which carries the daily risk of choking to death. But there is a further factor to take into account. As each adaptation begins to prevail, it is doing so in an environment which continues to change, and which varies from place to place. The human adaptations of upright walking and the big brain are universal to the species, but today's *Homo sapiens* does not live in a single environment.

The environment does not refer only to climate and topography. An individual's environment is the whole package of circumstances in which he or she lives – which clearly includes the weather and landscape but also extends to whether their life is rural or urban, whether their diet is good or bad, whether they are poor or rich, whether they have access to sophisticated, simple or non-existent medical care.

People who live in the developed economies of the West think of their environment as being dominated by cities, technology and economic resources. But the majority of the population of the globe does not live in that way. Most of the people of the world are rural, poor and without access to technology and material benefits.

The environment of someone living in a village in rural China is completely different to the environment of a New Yorker. A hunter–gatherer in the Kalahari presents a third set of circumstances, a citizen of Delhi a fourth. If evolution is about solving problems, then the problems of survival faced by these four humans are plainly widely at variance with each other.

This might suggest that evolution would select varying adaptations for varying parts of the modern population. But alongside the variety of the modern world is its fluidity. Against the huge variations in circumstances must be set the genetic interchange of the modern populations of the world. Genetic flow should lead to consistency of the species, but the variation in the environment should lead the other way. On this point Robert Foley remarks: "There is an enormous amount of movement and interbreeding and migration, which means that you don't end up with lots of different evolutionary directions, you end up with an extraordinary evolutionary complexity and almost an inertia.

"The future evolution of our species is going to be very hard for anyone to predict, because people live in such diverse situations."

Palaeoanthropologists are not soothsayers. They are not in the business of predicting the future, especially not in the modern scientific climate, with its great emphasis on careful observation of secure evidence.

On the other hand a better understanding of the past does give a much clearer picture of the present, particularly of the biological place occupied by humans. And that understanding makes it quite impossible to resist informed speculation about the future of the species. Unfortunately, its ultimate future is clear.

Humans are a species of primate, of which there are altogether about 200 species alive today. There may have been 4,000 to 5,000 species of primate in the past; all but today's crop of 200 are now extinct. "As a biologist," says Foley, "I would say the expectation is that our own species will go the way of any other species. It will become extinct. What we don't know, of course, is whether that extinction is going to be in 1,000 years or 10,000 years or one million years or ten million years. But extinction is the normal expectation for all species. And one can say that as we know the earth cannot last for ever, extinction is inevitable."
(Interview, October 1993.)

At Harvard University there used to be some graffiti on a wall, an anti-nuclear war protest. It was a stencilled image of a cockroach, with the legend underneath: "The Last Survivor." John Shea, who now teaches at the State University of New York, used to walk past the graffiti every day. "It shows you evolution will continue. It just doesn't necessarily mean it's going to involve humans.

"At some point the constellation of morphological features and genes that are within the modern human population will cease to occur. Every other species that we find in the fossil record has a limited span of time. That's not to say we won't have descendants, who will take some other form, but that will depend on the context in which those evolutionary changes occur. But one knows that evolution will continue."
(Interview, August 1993.)

We may not be here to witness it, but nothing can stop evolution.

What of the shorter term, the time between now and that point in the hopefully distant future when the species disappears? What changes can we expect, and what changes are perhaps already in the process of taking place?

Geneticists view evolution as changes in the frequencies of genes in a breeding population, and that is an active process. John Shea cites an example: "In the last 1,000 years Europeans have been very successful at reproducing themselves in areas where Europeans hadn't previously been. In the global sense of changes in gene frequencies, there's been a major evolutionary change. European genes have been reproduced far more broadly afield and in far greater numbers than they had in previous millennia."

Ways of living in the modern world.

Top left, a rice-growing village in China.
Above, the streets of Old Delhi in India.
Below, Masai man in East Africa in his traditional dress, alongside men in western dress.
Left, Manhattan Island, New York.

171

The diversity of modern humans.

Above, a Swedish girl in traditional headdress. Above right, a Bena Bena tribeswoman in Papua New Guinea wearing face paint, beads and feathers. Right, an Inuk woman from northern Canada.

172

Erik Trinkaus says that the sheer size, fluidity and momentum of the modern species militate against change. But he reinforces the importance of biology, and the fact that no species can step outside the process of evolution. No animal can ever be so clever that it can take its bat and ball away in order to play in a game with no genetic laws. "We can modify the environment, we can affect the fates of human populations, the fates of other species. But when push comes to shove, the baseline is our genetically determined anatomy and physiology.

"Despite all the medical advances that we have made and our tremendous ability to improve the quality of life of people who have congenital abnormalities, we have done very little to affect the overall genetic make-up of the human species.

"Mathematically it's virtually impossible to have any effect on the genetic direction of the human species unless you wipe out 99 percent of living humanity and start out with a very few select individuals."
(Interview, August 1993.)

Trinkaus estimates that over the past 10,000 years there have been virtually no genetically determined anatomical changes in the human species. Our anatomy evolved when we were hunter–gatherers, before we built cities and surrounded ourselves with all the trappings of what we call civilization.

However, before the last Ice Age there were some changes in our bodies. On the very heavy build of the Neandertals, Trinkaus says that it is not their shape which has to be explained but ours. We are the first hominids of slim build and weak muscles; we have, in Trinkaus's memorable phrase, "wimped out." We have become a little taller in the last few hundred years, but we have also become generally less strong than our predecessor species and probably more prone to some diseases as a result of a more sedentary, less active lifestyle. But these changes are not fundamental: the genetic basics of our physiology have not changed since they were put in place during, and before, the last Ice Age.

This means of course that at least that part of the world's population that lives in cities and moves around on wheels is not anatomically adapted to its current environment. We do not really fit it, even though we built much of it, supposedly to suit us. On the other hand, we are not likely to evolve a brand-new anatomy which would suit city life better, because of the overwhelming mobility of human genes constantly travelling through the world's population. If a single city were to be completely isolated from the whole of the rest of the world, it is possible to imagine that over time a new mutation might emerge and prevail: but that is simply not going to happen.

If there is not going to be a new city-dwelling hominid species, it is just possible that there could be a new hominid species in outer space. If humans continue to explore and then colonize other planets, the genetic barrier between another planet and earth could be sufficiently impenetrable to allow a new species to emerge. But for the moment, that possibility sits in the world of science fiction, and we are left to contemplate more realistic possibilities on earth.

"There is no question that the human species will continue to evolve," says Trinkaus, "but all I mean by that is that frequencies of genetically determined traits in populations will change gradually over generations. The rate of that biological evolution is going to be very small."

If major biological change is relatively unlikely, there is a greater chance of local adaptation. We know from the past that the human form is very plastic, that changes in appearance can take place in response to local circumstances within a relatively short period of time. It is possible to speculate, for example, about skin colour.

Pale skin is generally thought to have evolved in the northern hemisphere in order to allow the sun's rays to penetrate the skin and synthesize essential Vitamin D. Further south, where the sun is stronger, dark skin provides protection against the harmful effects of too

The shuttle Columbia about to touch down after an exploratory mission in earth orbit. Humans may now be able to fly beyond the earth's atmosphere, but they cannot control evolution.

much direct sunlight. There is also some evidence that light skin is less susceptible to injuries caused by the cold, such as frostbite.

Damage to the ozone layer could result in over-exposure to the sun in northern latitudes, with a consequent increase in skin cancer, which is already far more prevalent than was the case only a few years ago.

The evolutionary solution to this problem could be much darker skins where presently they are pale. The technological solution could of course be the universal use of sun-blocking cream and the abandonment of the idea of sun-bathing (a practice which has only been with us for a few decades.)

Local adaptations apart, the size, distribution and mobility of the human population restrict the chance of major genetic change, whose breeding ground is isolation. On the other hand, biological evolution cannot be stopped, predicted or controlled. According to Trinkaus, "The idea that we can somehow direct human evolution in the future, whether in a biological sense or a cultural sense, I think is pure fantasy. It's great for Star Trek, it's great for Star Wars, but it tells us very little about what is really going to happen to humans in the future."

The geneticists, relatively new to evolutionary science, want to know where the defining line sits which separates genetically controlled traits in humans from other elements in their make-up which are learned in each generation and inherited by way of passed-on knowledge rather than genes. Where do genetics stop and education begin?

Maryellen Ruvello says that up to this point most evolutionary genetic research has been to do with relationships and ancestry. The next big question is to find what is the genetic basis for culture, for language, for the other unique human qualities. In other words, the study of genetics having shown us how close we are to the chimpanzees, may now begin to tell us why we are different to them.

"I think there are probably genes for human culture, in the same way that chimpanzee culture has a genetic basis. We know that chimpanzee cultural variation can be quite great – some groups use tools, others do not. This is a heritage that we share in part with the chim-

panzees; clearly all humans use tools to a much greater extent than the chimpanzees, and I think there was probably a genetic basis for this."
(Interview, August 1993.)

So it may be shown in the future that biological evolution is to do not only with our anatomy but also with our behavior, or certain key elements of it. If there is a genetic basis for the use of tools, it follows from what we know that there might also be a genetic basis for the aspects of our lives which flow from the interaction of the development of the brain with our other activities: language, art and creativity.

This would finally break down the wall which separates anatomy from ideas. People who accept the basic idea of evolution sometimes feel a lingering loyalty to creationism when it comes to contemplation of the human soul, of human nature. It is as though the spirit came to us from some other source in the last 20,000 years or so. Something happened which is on a different level to mere physical changes; we acquired a loftier sense of self.

If one wish could be granted to Vincent Sarich, a leading molecular scientist at the University of California at Berkeley, it would be to break through this barrier in conventional thinking about human nature. "What we need to do is to try to bring an evolutionary perspective into looking at our behavior. Everything else is, I think, of relatively trivial consequence compared to this overriding problem. We do not want to accept the idea that our soul evolved. If I had one thing I could do, it would be to get that idea across."

Sunbathing may give a socially acceptable tan – but in fact it is extremely dangerous to expose pale skin to sunlight.

He points out that not only our anatomy evolved when we were hunter–gatherers. "Our feelings, our specifically human feelings, have been shaped in that environment – but we're not living in that environment any more. Population densities are probably a hundred times greater than they were, so we're having to interact in a more complex way with a much larger number of people than we're properly evolved to do.

"What human beings are learning to do is to adapt their hunter–gatherer developed psyches to a world in which they didn't evolve – and that's why the adaptation has been and continues to be bloody at times."
(Interview, August 1993.)

The control of territory and resources beyond the immediate group is not a usual ambition or practice among hunter–gatherer people. Warfare, according to Sarich, is a feature of human behavior which has arisen since the introduction of agriculture and very large increases in population, and is a consequence of the mismatch between circumstances and psyches. Maybe war will be seen in the future as "a little interruption in a generally far more peaceful existence between groups and within groups. We're still learning how to adapt to a world that is very different from the one that made us."

Warfare and violence have been a major preoccupation of evolutionary science. Sarich says that warfare is less than 10,000 years old and certainly there is nothing in the fossil or archaeological record to suggest it is any older than that. On the other hand personal violence is far from unknown among other primates, including chimpanzees, which will all behave aggressively in the interests of protecting resources, territory or their social rank. Some primate males practise infanticide when they take over a new troop in which there are already infants in order that they can sire their own offspring.

Humans may share some of the other primates' aggressive behavior; but humans have also developed culture and value systems which enable them either to shy away from violent behavior or to plan it more effectively and cause greater pain and destruction.

Technological progress has brought advanced weaponry into the reach of humans; we appear only to have conducted systematic warfare in the last 10,000 years.

Advanced human societies make up rules for daily living, and use symbols to convey their rules to each other.

Attitudes and behavior change constantly whether or not they are genetically controlled. Even the concept of free will, at the very heart of the human-value system, has arisen from the evolutionary process, rather than as something conferred on us by a magical outside force. Philip Lieberman believes that free will has a biological foundation related to the development of language.

Humans can modify their speech in any way they want. Sounds are not connected to emotion; we can choose to make a particular statement which is not relevant to, or which even contradicts, our emotional response. Modifying speech connects to modifying actions: "This means that we have a basis for belief that we can modify our actions in accord with our ultimate desires, and that, I think, is ultimately the basis of a concept like free will. Free will, compassion and altruism really depend on language." Human morality therefore has a biological basis; it flows from the adaptation of language.

Whether that means there is a gene for morality is another question, the answer to which is long in the future. One has to be careful about the notion of a gene for a particular type of behavior or attitude, if only because genes do not operate in isolation. They only work in the context of other genes and of the environment. But it is already very hard to identify any aspect of human life which cannot in one way or another be set in the framework of the evolutionary process.

There is a direct, observable relationship between culture and genes. Maryellen Ruvello quotes the example of the introduction of slash-and-burn agricultural techniques in West Africa. This involves the systematic destruction of areas of forest for cultivation, the principle being that it is cyclical; cultivated areas of land are allowed to return, on a rotational basis, to forest cover – always provided that enough food is being produced to meet the needs of the population. Slash-and-burn has been widely used in developing countries as a means of opening up agricultural land without permanently destroying the forests: the reality has been rather different, however, and it has had unforeseen consequences.

"Slash-and-burn in West Africa has led to an increased mosquito population," says Ruvello, "which has led to a greater exposure to malaria, and therefore led to increased selection of a gene for sickle cell – which offers a benefit against malaria." The sickle cell gene gives significant protection against malaria, vital in an area where in some parts of the population, 7 percent of children die of the disease.

Huge areas of rainforest have been destroyed all over the world to make way for agriculture, as here in Malaysia – but the open space has created fertile ground for malaria, one of the world's leading killer diseases. The sickle cell gene provides some protection against malaria but carries its own risk of fatal disease. The inset shows a Jamaican child with sickle cell anaemia being examined.

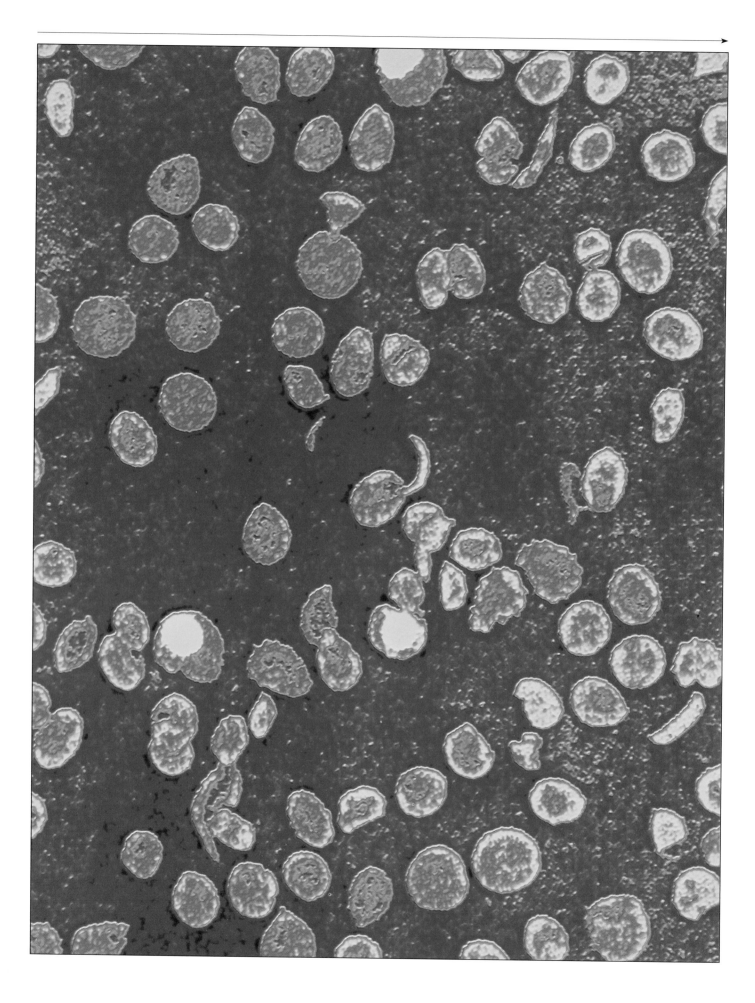

But the sickle cell gene itself can become an enormous problem if you happen to be the offspring of parents who both carry the gene. In that event, you have a one in four chance of dying from sickle cell anaemia. The more malaria there is in the population, the more people are likely to carry the sickle cell gene – and the more offspring are going to fall victim to anaemia. This chain of evolutionary events is set off by agriculture, an interesting example of the interaction of culture and natural selection; and of the fact that genes do not operate in isolation.

Free will may originate in evolution. That does not mean that our will is not free. The fact that it came to us indirectly by natural selection does not imply that all we have to do as humans is sit back and await further selective processes.

Bill Kimbel, at the Institute of Human Origins in California, returns to the question of our future. "On the one hand the fossil record is eloquent testimony to the fact that all species ultimately become extinct. On the other hand no species in the history of evolution has achieved such a complex cultural milieu as we have.

"We have in fact more or less sheltered ourselves from the forces of evolution in many significant ways. So the future of our species is very much in our own hands. The only question is whether we will ride out the perturbations in our environment that are ultimately of our own doing, and that will essentially determine our fate. We have the power to rectify that situation, but realistically speaking it doesn't appear that we're ready to do so."
(Interview, August 1993.)

OPPOSITE: *A blood smear shows the distorted red blood cells of sickle cell anaemia. The normal rounded cells can be seen alongside abnormal cells which have shrunk into a sickle shape. Sickle cell anaemia can be fatal in childhood, and no cure is yet known.*

BELOW: *Remote rainforest in West Kalimantan, Indonesia, being logged and cleared for new populations. Humans have a habit of drastically altering their environment for short-term benefit, the longer-term consequences of which may be anything but beneficial.*

Oxford Street, London: one of the busiest streets in the world and a symbol of so-called civilization. On the facing page, another sign of modern civilization, a deforested area of the Cameron Highlands in Malaysia.

The prospect of sophisticated humans failing to stop preventable ecological disaster preoccupies many scientists and is a bizarre irony of developed human culture. Elisabeth Vrba was brought up in Africa: "When I think about the future of our species, one of the things I would like to preserve for our children, and their children, is nature. And just on that score I feel pessimistic when I see how much of these ecosystems have become degraded and have disappeared, even during my lifetime.

"I especially know places in Africa where that is true. And I feel very strongly that a central concern for humans should be to think about the factors that govern population growth, so that technological improvements that humankind comes up with can be brought to bear to improve lives.

"If population growth constantly outraces the medical and other advances that we make, then misery and war will continue."

The population of the world in the middle of the eighteenth century was about 300 million. Its rate of increase at that time was very slow. Today, the population has reached about five and a half billion, and it is increasing at a little less than 2 percent a year. By the year 2050, the world's population could have nearly doubled, even if the present rate of increase slows down. Plainly, problems of over-population today could simply worsen in the future.

Some areas of the world, such as the Horn of Africa, are already unable to sustain all the people who live there, and the result is violence and suffering. Meanwhile, other living species are disappearing, in part due to our activities. Vrba says that at present we stand to lose up to thirty million species.

"This is a very disturbing thought because the diversity on earth is like a genetic bank. The investment of millions of years of evolution is sitting in those genes, and when these branches on the tree of life are lost, they are lost for ever.

Humans are not the only species to lay waste to their environment, although they do it on a wider scale than any other animal. Here, elephants use their tusks to remove bark from a tree, which will die.

"It looks as though the sun is still shining as before, but wherever you go, you know something major has gone, and it's gone for ever. That's not the world that I want to hand to my children's children."

(Interview, August 1993.)

Phillip Tobias is more optimistic, and he thinks that the study of the past does offer a basis to predict where we are going. He says there is a message of hope in our increased dependence on cultural and behavioral adaptation. "We are not changing very much in our anatomy any more. Perhaps we're losing our wisdom teeth, perhaps we're losing our little toes, but there's very little else that's changing in us other than a few minor changes in body size.

"But behaviorally, psychologically, spiritually, I think that the man and woman of the future is going to be very different from ourselves. I believe our evolution has entered upon a new plane. I believe we have the knowhow to develop answers to the problems we have created.

"And we have developed ethical systems. We have developed what we like to call principles, rules for the road of life, and we have developed certain aesthetic feelings. We like to see diversity in the environment, and, armed with these principles, I believe there is sufficient aliveness, alertness of people today to want to do something about it.

"I believe that mankind's knowledge and mankind's ethos are going to help him to overcome this tendency hell-bent to destroy our planet, to pollute it out of existence."

He is not altogether convinced by the argument that environmental destructiveness is a uniquely human foible. Other species lay waste to their surroundings in their own interests – big cats and hyenas may wipe out the local antelope population; elephants cause massive destruction to their habitats. But humans have acquired the means to do it more resolutely and more devastatingly than any other species, perhaps because of the sheer degree of human dominance.

"Maybe our mastery of the environment leaves something to be desired. Maybe we have got to temper and tone it down with a greater degree of love and respect for that which is around us."

(Interview, August 1993.)

British lawyers demonstrating several aspects of human evolution. They use language, specially encoded for their fellow lawyers; they enforce the rules of society, for the benefit of all; they wear special symbolic clothing to distinguish themselves from others; and they engage in daily competitive behaviour for their own interests.

His point is that cultural development has given humans the ability to feel that love and respect, and that will enable them to draw back before it is too late.

By contrast, Terrence Deacon thinks this ability to be responsible, kind and caring has another side. No other species anticipates death, and Deacon says the knowledge that consciousness will cease puts an emotional burden on humans which evolution has not equipped the brain to deal with.

"I think the fear of death drives many, many societies in directions that are destructive. It is also responsible for some of the most noble things about humans. The question is, which will win out?" We are capable of being generous and thoughtful, but we also know how to manipulate others, how to harm them: "We know how to cause them pain, we know about torture. The gift that has given us the most noble features of human mental life has also given us the most heinous features of mental life."
(Interview, August 1993.)

Of course no one knows what will happen in the future; the element of chance in evolution will make sure of that, if nothing else does. But as Phillip Tobias remarked, mammals are inquisitive creatures, they like to know what is going on. One of the attractive features of palaeoanthropology to an outsider is that it provides a pair of spectacles through which to peer at the private lives of our ancestors, and thus at ourselves.

The most significant moment in the history of the hominids was the time, not when the brain expanded, but when upright walking began. The new way of living of the upright hominids was the basis of everything which came later by way of new skills, brain power, technology and the eventual population of the world.

There is no such thing as being en route to humanity. Evolution does not allow the concept of a transitional species or a halfway house. At the time of its existence, each species lives in its own right and can only be seen in the context of its time, not of our retrospective desire to see our predecessors as creatures striving to become us.

Humans are at once ordinary and remarkable. Ordinary in that we are just another species of primate in the genus *Homo*. Remarkable in that we have acquired a package of characteristics and abilities not shared by any other creature, and giving us an extraordinary range of mental and physical powers.

An outsider who borrows palaeoanthropological spectacles for a short while is humbled and slightly unnerved to discover that the consciousness enabling him to learn something of human history and the language enabling him to express it are a product of the imperfect, short-term mechanisms of natural selection.

But above all, an appreciation of our history is an irreplaceable corrective for any ideas of superiority, whether of humans over other species or of one group of humans over another. After all, we all share an African ancestor.

FURTHER READING

This is by no means a comprehensive selection from the vast range of literature available on the subject of human evolution. But anyone who has been intrigued by the subjects touched upon in this book may want to find out more from authoritative sources – and each of the books listed here contains fascinating material.

Leslie Aiello and Christopher Dean, *An Introduction to Human Evolutionary Anatomy*, Academic Press, 1990

Paul G. Bahn and Jean Vertut, *Images of the Ice Age*, Facts on File, 1988

Jacob Bronowski, *The Ascent of Man*, BBC Books, 1973

Cambridge Encyclopedia of Human Evolution, Cambridge University Press, 1992

Dorothy Cheney and Robert Seyfarth, *How Monkeys See the World*, University of Chicago Press, 1990

Charles Darwin, *The Origin of Species*, Penguin Classics, 1985

Richard Dawkins, *The Blind Watchmaker*, Penguin Books, 1991

Richard Dawkins, *The Selfish Gene*, Oxford University Press, 1989

Jared Diamond, *The Rise and Fall of the Third Chimpanzee*, Vintage, 1992

Robert Foley, *Another Unique Species*, Longman Scientific and Technical, 1987

Robert Foley ed., *The Origins of Human Behaviour*, Unwin Hyman, 1991

Stephen Jay Gould, *Bully for Brontosaurus*, Penguin Books, 1992

Stephen Jay Gould, *Ever Since Darwin*, Penguin Books, 1991

Donald C. Johanson and Maitland E. Edey, *Lucy, the Beginnings of Mankind*, Penguin Books, 1990

Donald Johanson and James Shreeve, *Lucy's Child, The Discovery of a Human Ancestor*, Penguin Books, 1991

Richard Leakey and L. Jan Slikkerveer, *Man-Ape, Ape-Man*, Netherlands Foundation of Kenya Wildlife Service, 1993

Richard Leakey and Roger Lewin, *Origins Reconsidered*, Little Brown, 1992

Roger Lewin, *Bones of Contention*, Penguin Books, 1991

Roger Lewin, *Human Evolution, An Illustrated Introduction*, Third Edition, Blackwell Scientific Publications, 1993

Philip Lieberman, *The Biology and Evolution of Language*, Harvard University Press, 1984

Man's Place in Evolution, Second Edition, Natural History Museum Publications, Cambridge University Press, 1991

Jonathan Miller and Borin Van Loon, *Darwin for Beginners*, Icon Books, 1992

Dr David Norman, *Dinosaur!*, Boxtree, 1991

John Reader, *Missing Links: The Hunt for Earliest Man*, Penguin Books, 1990

Kathy D. Schick and Nicholas Toth, *Making Silent Stones Speak*, Simon & Schuster, 1993

Christopher Stringer and Clive Gamble, *In Search of the Neanderthals*, Thames and Hudson, 1993

Erik Trinkaus and Pat Shipman, *The Neandertals*, Jonanthan Cape, 1993

Randall White, *Dark Caves, Bright Visions*, American Museum of Natural History, 1986

INDEX

Numbers in *italics* represent illustrations

ACKNOWLEDGEMENTS

Paul Bahn and Jean Vertut, *Images of the Ice Age*, Facts on File Inc., 1988

Dr Jacob Bronowski, *The Ascent of Man*, BBC Enterprises Ltd and Science Horizons Inc., 1973

Cambridge Encyclopedia of Human Evolution, Cambridge University Press

Jared Diamond, *The Rise and Fall of the Third Chimpanzee*, Random House UK Ltd, 1992

Dr Robert Foley, *Another Unique Species*, Longman Scientific and Technical, 1987

Donald C. Johanson, *Lucy, the Beginnings of Mankind*, Penguin Books, 1990

Roger Lewin, *Bones of Contention*, Penguin Books, 1991

Roger Lewin, *Human Evolution, An Illustrated Introduction*, Blackwell Scientific Publications Ltd, 1993

John Reader, *Missing Links*, Penguin Books, 1990

PICTURE ACKNOWLEDGEMENTS

Dr Helmut Albrecht/Bruce Coleman 22 bottom, 61; **Brian and Cherry Alexander** 126 main pic, 172 bottom middle; **Ancient Art and Architecture Collection** 160, 161 bottom; **Trevor Barrett/Bruce Coleman** 91; **Erwin and Peggy Bauer/Bruce Coleman** 137, 168; **Tim Beddow/Science Photo Library** 66 main pic; **The Bridgeman Art Library/Bible Society, London** 132; with permission from **The British Library, London**, *Illustrated London News*, 14 Feb 1925, 73 top; © **The British Museum, London** 152, 153; **John Burton/Bruce Coleman** 31 top right; **R I M Campbell/Bruce Coleman** 184 inset; **Dr Colin Chumbley/Science Photo Library** 79 middle right; **CNRI/Science Photo Library** 66 inset; **David Coulson** 8; **Gerald Cubitt/Bruce Coleman** 111, 184 main pic; **Peter Davy/Bruce Coleman** 51, 55 bottom, 58, 64; **Ecoscene/Cooper** 28; **Ecoscene/W Lawler** 181; **Ecoscene/Sally Morgan** 178-79; **Ecoscene/Erik Scahffer** 183; **Ken Edward/Science Photo Library** 113 main pic; **ESA/PLI/Science Photo Library** 13; **Mary Evans Picture Library**, 144; **Sue Ford/Science Photo Library** 178 insert; **Christer Fredriksson/Unikexempler/Bruce Coleman** 31 bottom, 89; **Jeff Foot Productions/Bruce Coleman** 90; **GJLP/CNRI/Science Photo Library** 79 top right; **Robert Harding Picture Library** 29, 95, 96, 126 inset, 164, 166, 171 top left, top right, bottom left, 172 top left 174, 175, 176, 177, 182, 185; **Robert Harding Picture Library/Paul Freestone** 96; **James Holmes/Cellmark Dignostics/Science Photo Library** 115; **Carol Hughes/Bruce Coleman** 87; **Hulton Deutsch Picture Library** 27 top right, 36, 37; **Illustrated London News Picture Library**, *Illustrated London News*, 28 Dec 1912 Supplement IV & V, 69 main pic; © **Institute of Human Origins, Berkeley, USA**, photo by Nanci Kahn 41, 43; **Peter F R Jackson/Bruce Coleman** 92 top; **Jemma Jupp** 38, 73 bottom right, 110, 131 **Dr M P Kahl/Bruce Coleman** 57; **Keith Kent/Science Photo Library** 161 top; **The Kobal Collection** 146; **The Kobal Collection/20th Century Fox 1981, McFill Everett** 147; **collection Musee de l'Homme, Paris cl. A. Glory** 107, 158 top left, 159 bottom left, bottom right; **collection Musee de l'Homme, Paris cl. I Oster** 158 middle right; **Dr Norman Myers/Bruce Coleman** 35 inset; **Rita Nannim/Science Photo Library** 92 bottom; © **The National Museum of Kenya, Nairobi, Kenya** courtesy of Professor Bernard Wood, University of Liverpool 118 middle left; **Natural History Museum, London** 27 bottom, 69 inset, 70, 71, 77 top right, 78-79, 118 top left & bottom right, 122 top, 123 top, 127 top right, 148, 149, 150, 151 both, 156, 157, 162; **Oxford Molecular Biophysics/Science Photo Library** 108; **Alfred Pasieka/Science Photo Library** 180; **Dieter and Mary Plage/Bruce Coleman** 22 top; **Phillipe Plailly/Science Photo Library** 113 inset; **Popperfoto** 10, 76 top left, 165; **Popperfoto/Reuters** 14; **Yoel Rak, Tel Aviv University, Israel** 127 bottom; **John Reader/Science Photo Library** 18, 20, 33, 35 main pic, 40 both, 42 both, 44 both, 52 both, 67, 68, 72, 76 bottom right, 78 left, 81, 85 both, 88, 98, 99 both, 100, 116, 117 both, 123 bottom left, 124, 134, 140, 141, 145; **Hans Reinhard/Bruce Coleman** 84; **Frank Spooner Pictures** 158 bottom left, 159 top right; **Sporting Pictures UK Ltd** 23; **A. J. Stevens/Bruce Coleman** 55 top right; **Geoff Tompkinson/Science Photo Library** 112; **Tropix/M. Auckland** 122 bottom; **Tropix/D. Charlwood** 172 top right; **Tropix/J. Schmid** 171 bottom right; **Tropix/J. Woollard** 106, 167; courtesy of **Professor Emeritus Phillip V. Tobias, Director, Palaeo-anthropology Research Unit, University of Witwatersrand** 50, 76 top; © **1968 Turner Entertainment Co.** All rights reserved 34 bottom left; **Peter Ward/Bruce Coleman** 59; courtesy of **Professor Peter Wheeler, Liverpool John Moores University, Liverpool** 65.